S0-BXL-551

C. Truesdell

The Elements of Continuum Mechanics

Springer-Verlag New York Inc.

The Elements of Continuum Mechanics

C. Truesdell
Professor of Rational Mechanics
The Johns Hopkins University

Springer-Verlag New York Inc.

Lectures given in August–September 1965 for the
Department of Mechanical and Aerospace Engineering
Syracuse University, Syracuse, New York
Sponsored by the New York State Science and Technology Foundation.

Library of Congress Catalog Card Number 65-29241
Printed in the United States of America

Title No. 7329

Foreword

The lectures here reported were first delivered in August and September, 1965, for the Department of Mechanical and Aerospace Engineering at Syracuse University, New York under the sponsorship of the New York State Science and Technology Foundation. Lectures 1-6 and 22-23 are revised from a version prepared by Professor Kin N. Tong on the basis of a transcription of the lectures, kindly provided by Professor S. Eskinazi. The remainder of the text has been written out afresh from my own notes.

Much of the same ground was covered in my lectures to the Austra lian Mathematical Society's Summer Research Institute at Melbourne in January and February, 1966, and for the parts affected the text conforms to this latter presentation. I am grateful to Professors C.-C. Wang and K. N. Tong for criticism of the manuscript.

These lectures constitute a course, not a treatise. Names are attached to theorems justly, to the best of my knowledge, but are not intended to replace a history of the subject or references to the sources. The science presented here, based on the classical researches in continuum mechanics completed by 1905, in its present form is due, directly or indirectly and in the main, to Messrs. Rivlin, Ericksen, Noll, Toupin, and Coleman, with important special additions by Signorini, Reiner, Stoppelli, Richter, A. E. Green, Oldroyd, Shield, Adkins, Serrin, Gurtin, C.-C. Wang, R. Bowen, and others. More detailed treatment, with full references, for most but not all of the topics presented, may be found in The Non-Linear Field Theories of Mechanics.

C. T.

CONTENTS

Part I. General Principles

Part II. Fluids

Part III. Elastic Materials

Part IV. Fading Memory

Part V. Thermodynamics

Part VI. Statics

PART I. GENERAL PRINCIPLES

LECTURE 1: BODIES AND MOTIONS.

Introduction

Mechanics is a classical subject, and any good course necessarily contains much old material, often merely repeated, oftener recast in a form useful in a new context. The recent renascence of mechanics broadened and deepened it but rather elevated than annulled the older parts of the subject. The classical nature of mechanics reflects its greatness: Ever old and ever new, it continues to pour out for us understanding and application, linking a changing world to unchanged law. My lectures will include topics with which most listeners are already familiar, but I hope that in the greater generality and compactness of the modern approach, these classical areas will be welcome as a foundation and as a guide to method and viewpoint.

The primitive elements of mechanics are

1. Bodies.
2. Motions.
3. Forces.

These elements are governed by assumptions or laws which describe mechanics as a whole. The laws abstract the common features of all mechanical phenomena. The general principles are then illustrated by constitutive equations, which abstract the differences among bodies. Constitutive equations serve as models for different kinds of bodies. They define ideal materials, intended to represent aspects of the different behaviors of various materials in a physical world subject to simple and overriding laws.

In the last century and earlier, certain classical theories of materials were cultivated intensively and almost exclusively. The system of mass-points or rigid bodies, finite in number, is the object studied in the discipline called "analytical mechanics"; the perfect fluid and some aspects of the linearly viscous fluid exhaust "hydrodynamics" and "aerodynamics"; while the theories of the perfectly flexible line, the membrane, the perfectly elastic solid, the infinitesimally visco-elastic material, and the perfectly plastic material are defined by other special constitutive equations. The concentration of all mechanical studies upon these theories, in fact special and even degenerate, caused them to become identified in the minds of many persons with mechanics as a whole. Indeed, in many universities today the student is considered a master of "classical mechanics" if he has a nodding acquaintance with the formalism of analytical dynamics. Just as bad, many engineering schools are now disguising by the name "continuum mechanics" a loose gruel of bits of classical elasticity floating at random in classical fluids wastes.

The new mechanics generalizes the old in four ways. First, a body is no longer necessarily regarded as an assembly of points; it may have microstructure. Second, there is much study of motions that are not always smooth; shock waves and other discontinuities, penetration of one body into another, and diffusion are allowed. Third, simple forces no longer exhaust the actions to which bodies are subject; couples, couple stresses, and higher moments have been introduced. Fourth and most important, a vast increase in generality has come in the constitutive equations. Not only is the variety of ideal materials become enormously greater, but also we now have efficient methods of describing and analyzing whole classes of theories at one time. Thus we look upon mechanics today, not as the budget of old rules common in teaching the subject until the last decade, but rather as an expanding framework for theories and experiments on the behavior of different sorts of materials. If you like, you may regard the new mechanics as a theory of materials in the same sense that geometry is a theory of figures. Just as we now study geometry as a whole, not merely the properties of a few special figures such as triangles and circles, so also we study mechanics as a whole, not merely the special materials whose properties fill the classical textbooks.

In these lectures I will present a conservative introduction to modern mechanical theory and to modern thought in mechanics. I will lay out what seems to me to be the core of what one must know about mechanics if he is either to work as a theorist in it or to apply it intelligently to experiments. I do not seek maximum generality. Rather, I will try to set forth those recent developments that have permanence, those, I can say with confidence, that will continue to be as valuable twenty years or fifty years from now as they are today. In particular, I shall nor present anything more general about bodies, motions, and forces, remaining content to deal with these in their classical forms, though not necessarily with all the old words and symbols. The main subject of these lectures is a modern theory of constitutive equations, but here, too, rather than seeking maximum generality I will explain the core of modern thought, the simplest, most elegant, and most illuminating parts of the recent work.

While mechanics, like geometry, ought to be developed consecutively from a set of axioms, such axioms are not yet settled. Axiomatics are never studied fruitfully until a conflict arises and a subject divides. In classical geometry, it was the creation of non-Euclidean geometries that made plain the need for modern axioms for Euclidean geometry. Mechanics has just recently come to be generalized on the one hand and made more precise on the other. Generality and precision sometimes oppose one another. NOLL has set up and studied a set of axioms for mechanics, but it is not general enough for the whole field today. While NOLL'S system would suffice as a basis for all I shall present in these lectures, it is preferable to let him speak for himself through the notes of his recent course ar Bressanone, which will soon be available. I shall give merely a descriptive treatment of the foundations, since I wish to pass as quickly as possible into the theory proper.

Finally, I am presuming that you who are listening are already interested in the subject, that you are theorists who know that good mathematics is to be found here, or experimenters who know the phenomena and are aware of the recent successes of the theory, or persons from other fields who wish to know what this one is. In particular, I am not attempting to proselytize. There are already too many people working in this field. What the foundations of mechanics need is not more people to do research on them but instead more people to learn and absorb and master and apply what has been done. Furthermore, I see no need to defend the applicability of modern mechanics to experiment or to problem-solving, since it is now obvious. Much of the older work claimed to solve a lot of problems without ever stating what those problems were or finding out what theory was at issue. This sort of stuff is now lying deep in the trash bin. When you have the right basic concepts, the solving of problems becomes either easy or impossible. When you understand what the problem is, the solution is far easier to find, if it can be gotten at all. Often one learns that the original problem was not the right one to attack in the first place.

Geometry.

A place, denoted by $\mathbf{x}$ or $\mathbf{y}$, is a point in Euclidean 3-dimensional space. Generally the results we shall derive hold just as well in n-dimensional space, but in all cases where it makes a difference, n is taken as 3. The time t is a real variable, $-\infty < t < \infty$.

A vector $\mathbf{v}$, $\mathbf{w}$, ... is a transformation of space onto itself. The operation of transformation is indicated by a + sign:

$$\mathbf{y} = \mathbf{x} + \mathbf{v}, \quad \mathbf{v} = \mathbf{y} - \mathbf{x},$$

where the second equation, defining the difference of two points, is merely an alternative for the first.

A tensor $\mathbf{T}$ is a linear transformation of the space of all vectors. The operation of transformation is written multiplicatively:

$$\mathbf{w} = \mathbf{T}\mathbf{v} \, .$$

Components of vectors and tensors will be needed rarely. The notations $v^k, v_k, T^{km}, T^k{}_m, T_k{}^m, T_{km}$ are standard; for orthogonal co-ordinates it is sometimes helpful to introduce the physical components $v\langle k\rangle$, $T\langle km\rangle$, which are components with respect to an orthonormal frame tangent to the co-ordinate curves at the point in question. $\mathbf{1}$ is the identity tensor. Its components of the above four types are g^{km}, δ^k_m, δ^m_k and g_{km}, respectively. The co-ordinates of $\mathbf{x}$, in any system, are denoted in these lectures by x_1, x_2, x_3.

The scalar, tensor, and outer products of two vectors $\mathbf{v}$ and $\mathbf{w}$ are denoted by $\mathbf{v} \cdot \mathbf{w}$, $\mathbf{v} \otimes \mathbf{w}$, and $\mathbf{v} \wedge \mathbf{w}$. The composition or product of the tensors $\mathbf{S}$ and $\mathbf{T}$ is written $\mathbf{T}\mathbf{S}$. The determinant, trace, transpose, and magnitude of $\mathbf{T}$ are written $\det \mathbf{T}$, $\mathrm{tr}\mathbf{T}$, $\mathbf{T}^T$, $|\mathbf{T}|$, where $|\mathbf{T}| = \sqrt{(\mathrm{tr}\mathbf{T}\mathbf{T}^T)}$.

A tensor $\mathbf{Q}$ is orthogonal if $\mathbf{Q}\mathbf{Q}^T = \mathbf{1}$. A tensor $\mathbf{T}$ is symmetric if $\mathbf{T} = \mathbf{T}^T$, skew if $\mathbf{T} = -\mathbf{T}^T$.

The gradient of a vector field $\underset{\sim}{v}(\underset{\sim}{x})$ is written as $\nabla\underset{\sim}{v}$ and is defined by

$$(\nabla \underset{\sim}{v})\underset{\sim}{a} = \lim_{s\to 0} \frac{1}{s} [\underset{\sim}{v}(\underset{\sim}{x} + s\underset{\sim}{a}) - \underset{\sim}{v}(\underset{\sim}{x})]$$

for all vectors $\underset{\sim}{a}$. The covariant components of $(\nabla\underset{\sim}{v})\underset{\sim}{a}$ are $v_{k,m}a^m$, where the comma denotes the covariant derivative. The gradient of a tensor may similarly be defined by regarding the tensor as a vector in a space of higher dimension.

Bodies, Configurations, Motions.

A body $\mathcal{B}$ is a manifold of particles, denoted by X. These particles are primitive elements of mechanics in the sense that numbers are primitive elements in analysis. Bodies are sets of particles. In continuum mechanics the body manifold is assumed to be smooth, that is, a diffeomorph of a domain in Euclidean space. Thus, by assumption, the particles X can be set into one-to-one correspondence with triples of real numbers X_1, X_2, X_3, where the X_α run over a finite set of closed intervals. Such triples are sometimes called "intrinsic co-ordinates" of the particles, but I shall not need to use them explicitly. The mapping from the manifold to the domain is assumed differentiable as many times as desired, usually two or three times, without further mention. The level of rigor here is that common in classical differential geometry, where the context makes it clear how many derivatives are assumed to exist.

The body $\mathcal{B}$ is assumed also to be a σ-finite measure space with non-negative measure $M(\mathcal{p})$ defined over a σ-ring of subsets $\mathcal{p}$, which are called the parts of $\mathcal{B}$. Henceforth any subset of $\mathcal{B}$ to which we shall refer will be assumed to be measurable[1]. The measure $M(\mathcal{p})$ is called the mass distribution in $\mathcal{B}$. It is assigned, once and for all, to any body we shall consider.

Bodies are available to us only in their configurations, the regions they happen to occupy in Euclidean space at some time. These configurations are not to be confused with the bodies themselves. In analytical dynamics,

---

[1] The open sets of $\mathcal{B}$ are assumed to be members of the σ-ring of sets. The members of the smallest σ-ring containing the open sets are called the Borel sets of $\mathcal{B}$. In many cases only Borel sets need be thought of when we speak of "parts" of a body.

the masses are discrete and hence stand in one-to-one correspondence with the numbers 1, 2, ... , n . Nobody ever confuses the sixth particle with the number 6, or with the place the sixth particle happens to occupy at some time. The number 6 is merely a label attached to the particle, and any other would do just as well. Similarly, in continuum mechanics a body $\mathcal{B}$ may occupy infinitely many different regions, none of which is to be confused with the body itself.

The two fundamental properties of a continuum $\mathcal{B}$ are, then:

1. $\mathcal{B}$ consists of a finite number of parts which can be mapped onto cubes in Euclidean space.
2. $\mathcal{B}$ is a measure space.

These are the simplest, most fundamental things we can say in summary of our experience with actual bodies and with the older theories designed to represent physical bodies in mathematical terms.

We refrain from imputing any particular geometry to the body $\mathcal{B}$, because we have no basis for doing so, since a body is never encountered directly. While the above two assumptions imply some geometric structure in $\mathcal{B}$, it seems to me closer to physical experience and hence preferable to turn instead to the properties of $\mathcal{B}$ in its configurations, for it is the configurations that are available to us.

Since a body can be mapped smoothly onto a domain, it can be mapped smoothly onto any topological equivalent of that domain. A sequence of such mappings,

$$\underset{\sim}{x} = \underset{\sim}{\chi}(X,t) , \qquad (1.1)$$

is called a motion of $\mathcal{B}$. Here X is a particle, t is the time ($-\infty < t < \infty$), and $\underset{\sim}{x}$ is a place in Euclidean space. The value of $\underset{\sim}{\chi}$ is the place $\underset{\sim}{x}$ that the particle X comes to occupy at the time t. The notations $\mathcal{B}_{\underset{\sim}{\chi}}$ and $\mathcal{P}_{\underset{\sim}{\chi}}$ will indicate the configurations of $\mathcal{B}$ and $\mathcal{P}$ at the time t. We shall consider only motions that are smooth in the further sense that $\underset{\sim}{\chi}$ is differentiable with respect to t as many times as may be needed. The <u>velocity</u> $\dot{\underset{\sim}{x}}$, <u>acceleration</u> $\ddot{\underset{\sim}{x}}$, and n^{th} <u>acceleration</u> $\overset{(n)}{\underset{\sim}{x}}$ of the particle X at the time t are defined as usual:

$$\dot{\tilde{x}} \equiv \partial_t \, \tilde{\chi}(X,t),$$

$$\ddot{\tilde{x}} \equiv \partial_t^2 \, \tilde{\chi}(X,t), \qquad (1.2)$$

$$\overset{(n)}{\tilde{x}} \equiv \partial_t^n \, \tilde{\chi}(X,t),$$

so that

$$\dot{\tilde{x}} = \overset{(1)}{\tilde{x}}, \; \ddot{\tilde{x}} = \overset{(2)}{\tilde{x}} \; .$$

Mass-density.

$\mathcal{B}$ is a continuum since it is a topological image of a domain. However, the mass distribution $M(p)$ is so far left arbitrary and might be discrete, or partially so. Of primary interest in continuum mechanics are masses which are absolutely continuous functions of volume. By volume I mean Borel measure[1] in the space of configurations, Euclidean space. To assume that $M(\cdot)$ is absolutely continuous is to assume that in every configuration, a part p of sufficiently small volume has arbitrarily small mass. Thus, formally, concentrated masses are excluded, and analytical dynamics will not emerge directly as a special case of continuum mechanics.

By the Radon-Nikodym theorem, the mass of p may be expressed in terms of a mass-density $\rho_{\tilde{\chi}}$:

$$M(p) = \int_{p_{\tilde{\chi}}} \rho_{\tilde{\chi}} dv \qquad (1.3)$$

where the integral is a Lebesgue integral. Clearly the function $\rho_{\tilde{\chi}}$ depends on $\tilde{\chi}$, and, as indicated, the integration is carried out over the configuration $p_{\tilde{\chi}}$ of p.

The existence of a mass-density expresses a relation between the abstract body $\mathcal{B}$ and the configuration $\mathcal{B}_{\chi}$ it occupies. For a suitably chosen sequence of parts p_k, nested so that $p_{k+1} \subset p_k$, that all the p_k have but the single point $\tilde{x}$ in common, and that the volume $v(p_k)$ approaches 0 as $k \to \infty$, the density is the ultimate ratio of mass to volume:

---

[1] The field of Lebesgue-measurable sets contains certain null sets that are not Borel sets, but these seem not to be of interest in continuum mechanics, so Borel measure suffices.

$$\rho_{\chi}(\mathbf{x},t) = \lim_{\kappa\to\infty} \frac{M(p_\kappa)}{V(p_\kappa)} \tag{1.4}$$

To find the relation between the mass-densities corresponding to different configurations of the same part p, we begin with the formulae

$$M(p) = \int_{p_{\chi_1}} \rho_{\chi_1} \, dv = \int_{p_{\chi_2}} \rho_{\chi_2} \, dv \ . \tag{1.5}$$

If we let λ stand for the mapping that carries χ_1 into χ_2 and write J for the absolute value of its Jacobian determinant:

$$J \equiv |\det \nabla \lambda| \ , \tag{1.6}$$

then

$$\int_{p_{\chi_1}} \rho_{\chi_1} \, dv = \int_{p_{\chi_1}} \rho_{\chi_2} J dv, \tag{1.7}$$

for every part p. Hence follows an equation relating the two densities:

$$\rho_{\chi_2} J = \rho_{\chi_1} \ . \tag{1.8}$$

This equation shows that the density in any one configuration determines the densities in all others.

Reference Configuration.

Often it is convenient to select one particular configuration and refer everything concerning the body to that configuration, which need be only a possible one, not one ever occupied by the body. Let κ be such a configuration. Then the mapping

$$\mathbf{X} = \kappa(X) \tag{1.9}$$

gives the place $\mathbf{X}$ occupied by the particle X in the configuration κ. Since this mapping is smooth, by assumption,

$$X = \kappa^{-1}(\mathbf{X}) \ . \tag{1.10}$$

Hence the motion (1.1) may be written in the form

$$\underset{\sim}{x} = \underset{\sim}{\chi}[\underset{\sim}{\kappa}^{-1}(\underset{\sim}{X}) ,t] \equiv \underset{\sim}{\chi}_{\underset{\sim}{\kappa}}(\underset{\sim}{X},t) . \tag{1.11}$$

In the description furnished by this equation, the motion is a sequence of mappings of the <u>reference configuration</u> $\underset{\sim}{\kappa}$ onto the actual configurations $\underset{\sim}{\chi}$. Thus the motion is visualized as mapping of parts of space onto parts of space. A reference configuration is introduced so as to allow us to employ the apparatus of Euclidean geometry.

The choice of reference configuration, like the choice of a co-ordinate system, is arbitrary. The reference configuration, which may be any smooth image of the body, need not even be a configuration ever occupied by the body in the course of its motion. For each different $\underset{\sim}{\kappa}$, a different function $\underset{\sim}{\chi}_{\underset{\sim}{\kappa}}$ results in (1.1). Thus <u>one</u> motion of the body is represented by <u>infinitely many</u> different motions of parts of space, one for each choice of $\underset{\sim}{\kappa}$. For some choice of $\underset{\sim}{\kappa}$, we may get a particularly simple description, just as in geometry one choice of co-ordinates may lead to a simple equation for a particular figure, but the reference configuration itself has nothing to do with such motions as it may be used to describe, just as the co-ordinate system has nothing to do with geometrical figures themselves. A reference configuration is introduced so as to allow the use of mathematical apparatus familiar in other contexts. Again there is an analogy to co-ordinate geometry, where co-ordinates are introduced, not because they are natural or germane to geometry, but because they allow the familiar apparatus of algebra to be applied at once.

<u>Descriptions of Motion</u>.

There are four methods of describing the motion of a body: the material, the referential, the spatial, and the relative. Because of our hypotheses of smoothness, all are equivalent.

In the <u>material description</u> we deal directly with the abstract particles X. This description corresponds to the only one used in analytical dynamics, where we always speak of the first, second, ..., $n^{\underline{th}}$ masses. To be precise, there we should say, "the mass-point X_i whose mass is m_i," but commonly this expression is abbreviated to "the mass i" or "the body m_i", etc. In a continuous body $\mathcal{B}$ there are infinitely many particles X. In the material description, the independent variables are X and t, the particle and the time. While the material description is the most natural in concept, it was not mentioned in continuum mechanics until 1951 and is still used little. For some time the term "material description" was used to denote another and older description often confused with it, the description to which we turn next.

The referential description employs a reference configuration. In the mid-eighteenth century Euler introduced the description which hydrodynamicists still call "Lagrangean". This is a particular referential description, in which the Cartesian co-ordinates of the position $\underset{\sim}{X}$ of the particle X at the time $t = 0$ are used as a label for the particle X. It was recognized that such labelling by initial co-ordinates is arbitrary, and writers on the foundations of hydrodynamics have often mentioned that the results must be and are independent of the choice of the initial time, and some have remarked that the parameters of any triple system of surfaces moving with the material will do just as well. The referential description, taking $\underset{\sim}{X}$ and t as independent variables, includes all these possibilities. The referential description, in some form, is always used in classical elasticity theory, and the best studies of the foundations of classical hydrodynamics have employed it almost without fail. It is the description commonly used in modern works on continuum mechanics, and we shall use it in this course. We must always bear in mind that the choice of $\underset{\sim}{\kappa}$ is ours, that $\underset{\sim}{\kappa}$ is merely some configuration the body might occupy, and that physically significant results must be independent of the choice of $\underset{\sim}{\kappa}$. Any motion has infinitely many differential referential descriptions, equally valid.

In the spatial description, attention is focused on the present configuration of the body, the region of space currently occupied by the body. This description, which was introduced by D'Alembert, is called "Eulerian" by the hydrodynamicists. The place $\underset{\sim}{x}$ and the time t are taken as independent variables. Since (1.1) is invertible,

$$X = \underset{\sim}{\chi}^{-1}(\underset{\sim}{x},t) , \qquad (1.12)$$

any function $f(X,t)$ may be replaced by a function of $\underset{\sim}{x}$ and t:

$$f(X,t) = f[\underset{\sim}{\chi}^{-1}(\underset{\sim}{x},t), t] \equiv F(\underset{\sim}{x},t) . \qquad (1.13)$$

The function F, moreover, is unique. Thus, while there are infinitely many referential descriptions of a given motion, there is only one spatial description. With the spatial description, we watch what is occurring in a fixed region of space as time goes on. This description seems perfectly suited to studies of fluids, where often a rapidly deforming mass comes from no-one knows where and goes no-one knows whither, so that we may prefer to consider what is going on here and now. However convenient kinematically, the spatial description is awkward for questions of principle in mechanics, since in fact what is happening to the body, not to the region of space occupied by the body, enters the laws of dynamics. This difficulty is reflected by the mathematical gymnastics writers of textbooks on aerodynamics often go through in order to get formulae which are easy and obvious in the material or referential descriptions.

According to (1.13), the value of any smooth function of the particles of $\mathcal{B}$ at time t is given also by a field defined over the configuration $\mathcal{B}_{\chi}$ at time t. In this way, for example, we obtain from (1.2) the velocity field and the acceleration field:

$$\dot{\mathbf{x}} = \dot{\mathbf{x}}(\mathbf{x},t) \ , \ \ddot{\mathbf{x}} = \ddot{\mathbf{x}}(\mathbf{x},t) \ . \tag{1.14}$$

Here we have written $\dot{\mathbf{x}}$ and $\ddot{\mathbf{x}}$ in two senses each: as the field functions and as their values.

Deformation Gradient.

The gradient of (1.11) is called the deformation gradient $\mathbf{F}$:

$$\mathbf{F} \equiv \mathbf{F}_{\kappa}(\mathbf{X},t) \equiv \nabla\chi_{\kappa}(\mathbf{X},t) \ . \tag{1.15}$$

It is the linear approximation to the mapping χ_{κ}. More precisely, we should call it the gradient of the deformation from κ to χ, but when, as is usual, a single reference configuration κ is laid down and kept fixed, no confusion should result from failure to remind ourselves that the very concept of a deformation gradient presumes use of a reference configuration. If, as we may, we select independently any curvilinear co-ordinates X_{α} and x_m in the reference configuration and the present configuration, respectively, so that the motion (1.11) assumes the co-ordinate form

$$x_m = \chi_{\kappa m}(X_1,X_2,X_3,t) \ , \quad m=1,2,3, \tag{1.16}$$

then the components of $\underset{\sim}{F}$ are simply the nine partial derivatives of the functions $\chi_{\underset{\sim}{\kappa}m}$ with respect to the X_α, viz

$$F^m_{\ \alpha} = x^m_{\ ,\alpha} \equiv \partial_{X_\alpha} [\chi_{\underset{\sim}{\kappa}m} (X_1, X_2, X_3, t)]$$

$$m = 1, 2, 3, \quad \alpha = 1, 2, 3 . \qquad (1.17)$$

Going back to (1.8), we derive EULER'S referential equation for the density:

$$\rho J = \rho_{\underset{\sim}{\kappa}} , \qquad (1.18)$$

where ρ is written for $\rho_{\underset{\sim}{\chi}}$, the mass density in the present configuration, and where

$$J \equiv |\det \underset{\sim}{F}| . \qquad (1.19)$$

Henceforth J will be used in the sense just defined rather than in the more general one expressed by (1.6). Since $\underset{\sim}{\chi}_{\underset{\sim}{\kappa}}$ is invertible, $\det \underset{\sim}{F}$ is of one sign for all $\underset{\sim}{X}$ and t, for a given reference configuration $\underset{\sim}{\kappa}$. While (1.18) is often called "the Lagrangean equation of continuity," that name is misleading, since if the motion is smooth, (1.18) holds, but if the motion is not smooth, generally J cannot be defined at all, so (1.18) becomes impossible to consider as a condition. The proper way to interpret (1.18) is to regard it as a condition giving the present density ρ, once the density $\rho_{\underset{\sim}{\kappa}}$ in the reference configuration is known

Exercise 1.1. By using the formula for differentiating a determinant, derive the following identity of EULER:

$$\dot{J} = J \operatorname{div} \dot{\underset{\sim}{x}} , \qquad (1.20)$$

where the dot denotes the time-derivative of $J(\underset{\sim}{X},t)$ and where $\operatorname{div} \dot{\underset{\sim}{x}}$ is the divergence of the velocity field $(1.14)_1$.

If we differentiate (1.18) with respect to time, regarding all quantities in it as functions of $\underset{\sim}{X}$ and t, and then use (1.20), we obtain D'ALEMBERT and EULER'S spatial equation for the density:

$$\dot{\rho} + \rho \operatorname{div} \dot{\mathbf{x}} = 0 \ . \tag{1.21}$$

This equation has exactly the same meaning as (1.18), which, conversely, may be gotten from it by integration.

Exercise 1.2. A motion is called isochoric if the volume $V(\mathcal{P})$ of each part $\mathcal{P}$ of the body remains constant in time. Show that any one of the following three equations is a necessary and sufficient condition for isochoric motion:

$$\operatorname{div} \dot{\mathbf{x}} = 0 \ , \ \rho = \rho_{\mathbf{K}} \ , \quad J = 1 \ . \tag{1.22}$$

Material Time Rates and Gradients in the Spatial Description.

In continuum mechanics the need to distinguish a vast number of quantities often deprives us of the luxury of using for a function a symbol different from that for its value, as logically we ought to do. If two functions of different variables have the same value and if both are denoted by that value, when we come to differentiate it is not clear which function is intended. The distinction, which of course is essential, is made by introducing different symbols for the differential operators. Henceforth

$$\dot{f} \quad \text{and} \quad \nabla f$$

will be used to denote the partial time derivative and the gradient of the function $G(\mathbf{X},t)$ such that

$$f = G(\mathbf{X},t) \ , \tag{1.23}$$

while

$$\partial_t f \quad \text{and} \quad \operatorname{grad} f$$

shall denote the time derivative and the gradient of the function $g(\mathbf{x},t)$ that has the same value as G, namely,

$$f = g(\mathbf{x},t) \ . \tag{1.24}$$

Since $\mathbf{x} = \chi_{\mathbf{K}}(\mathbf{X},t)$, application of the chain rule to the equation $G(\mathbf{X},t) = g(\mathbf{x},t)$ yields the classical formula of EULER:

$$f = \partial_t f + (\operatorname{grad} f)\dot{\mathbf{x}} \ . \qquad (1.25)$$

In particular, the acceleration $\ddot{\mathbf{x}}$ is calculated from the velocity field $\dot{\mathbf{x}}(\mathbf{x},t)$ by the formula

$$\ddot{\mathbf{x}} = \partial_t\dot{\mathbf{x}} + (\operatorname{grad} \dot{\mathbf{x}})\dot{\mathbf{x}} \ . \qquad (1.26)$$

Likewise,

$$\nabla f = \mathbf{F}^T \operatorname{grad} f \ . \qquad (1.27)$$

The notations "div" and "curl" will be used only in the spatial description, and superimposed (n) shall stand for n superimposed dots. The notation "Div" shall stand for the divergence formed from ∇ rather than from grad.

<u>Change of Reference Configuration</u>.

Let the same motion (1.1) be described in terms of two different reference configurations, $\boldsymbol{\kappa}_1$ and $\boldsymbol{\kappa}_2$:

$$\mathbf{x} = \boldsymbol{\chi}_{\kappa_1}(\mathbf{X},t) = \boldsymbol{\chi}_{\kappa_2}(\mathbf{X},t) \ . \qquad (1.28)$$

The deformation gradients $\mathbf{F}_1$ and $\mathbf{F}_2$ at $\mathbf{X},t$ are of course generally different. Let $\mathbf{X}_1$ and $\mathbf{X}_2$ denote the positions of X in $\boldsymbol{\kappa}_1$ and $\boldsymbol{\kappa}_2$:

$$\mathbf{X}_1 = \boldsymbol{\kappa}_1(X) \ , \qquad \mathbf{X}_2 = \boldsymbol{\kappa}_2(X) \ . \qquad (1.29)$$

Then

$$\mathbf{X}_2 = \boldsymbol{\kappa}_2[\boldsymbol{\kappa}_1^{-1}(\mathbf{X}_1)] \equiv \boldsymbol{\lambda}(\mathbf{X}_1) \ , \qquad (1.30)$$

say. The deformation from $\boldsymbol{\kappa}_1$ to $\boldsymbol{\chi}$ can be effected in two ways: either straight off by use of $\boldsymbol{\chi}_{\kappa_1}$, or by using $\boldsymbol{\lambda}$ to get to $\boldsymbol{\kappa}_2$ and then using $\boldsymbol{\chi}_{\kappa_2}$ to get to $\boldsymbol{\chi}$. If a circle denotes the succession of mappings, then,

$$\boldsymbol{\chi}_{\kappa_1} = \boldsymbol{\chi}_{\kappa_2} \circ \boldsymbol{\lambda} \ . \qquad (1.31)$$

Since this relation holds among the three mappings, their linear approximations, the gradients, are related in the same way:

$$\mathbf{F}_1 = \mathbf{F}_2\mathbf{P} \ , \qquad (1.32)$$

where

$$\underset{\sim}{P} \equiv \nabla \underset{\sim}{\lambda} \ . \tag{1.33}$$

Of course, the relation (1.32) expresses the "chain rule" of differential calculus.

Current Configuration as Reference.

To serve as a reference, a configuration need only be a diffeomorph of the body. Thus far, we have employed a reference configuration fixed in time, but we could just as well use a varying one. In this way a given motion may be described in terms of any other. The only variable reference configuration really useful in this way is the present one. If we take it as reference, we describe the past and future as they seem to an observer fixed to the particle X now at the place x. The corresponding description is called relative.

To see how such a description is constructed, consider the configurations of $\mathcal{B}$ at the two times t and τ:

$$\underset{\sim}{\xi} = \underset{\sim}{\chi}(X,\tau) \ , \tag{1.34}$$

$$\underset{\sim}{x} = \underset{\sim}{\chi}(X,t) \ .$$

That is, $\underset{\sim}{\xi}$ is the place occupied at time τ by the particle that at time t occupies $\underset{\sim}{x}$:

$$\underset{\sim}{\xi} = \underset{\sim}{\chi}[\underset{\sim}{\chi}^{-1}(\underset{\sim}{x},t) \ , \ \tau] \ ,$$

$$\equiv \underset{\sim}{\chi}_t(\underset{\sim}{x}, \ \tau) \ , \tag{1.35}$$

say. The function $\underset{\sim}{\chi}_t$ just defined is called the relative deformation function.

Sometimes we shall wish to calculate the relative deformation functions when the motion is given to us only through the spatial description of the velocity field:

$$\dot{\underset{\sim}{x}} = \dot{\underset{\sim}{x}}(\underset{\sim}{x},t) \ . \tag{1.36}$$

By $(1.35)_1$,

$$\partial_{\tau}\underset{\sim}{\xi} = \dot{\underset{\sim}{x}}(\underset{\sim}{\xi},\tau) \ . \tag{1.37}$$

Since the right-hand side is a given function, we thus have a differential equation to integrate. The initial condition to be satisfied by the integral $\boldsymbol{\xi} = \boldsymbol{\chi}_t(\mathbf{x},t)$ is

$$\boldsymbol{\xi}\Big|_{\tau=t} = \boldsymbol{\chi}_t(\mathbf{x},t) = \mathbf{x} \ . \tag{1.38}$$

When the motion is described by (1.35), we shall use a subscript t to denote quantities derived from $\boldsymbol{\chi}_t$. Thus $\mathbf{F}_t$, defined by

$$\mathbf{F}_t \equiv \mathbf{F}_t(\tau) \equiv \operatorname{grad} \boldsymbol{\chi}_t \ , \tag{1.39}$$

is the <u>relative deformation gradient</u>.

By (1.32), at $\mathbf{X}$

$$\mathbf{F}(\tau) = \mathbf{F}_t(\tau)\, \mathbf{F}(t) \ . \tag{1.40}$$

As the fixed reference configuration with respect to which $F(\tau)$ and $F(t)$ are taken we may select the configuration occupied by the body at time t'. Then (1.40) yields

$$\mathbf{F}_{t'}(\tau) = \mathbf{F}_t(\tau)\, \mathbf{F}_{t'}(t) \ . \tag{1.41}$$

Of course

$$\mathbf{F}_t(t) = \mathbf{1} \ . \tag{1.42}$$

LECTURE 2: KINEMATICS. CHANGES OF FRAME.

Stretch and Rotation.

Since the motion χ_{κ} is continuous, $\mathbf{F}$ is non-singular, so the polar decomposition theorem of CAUCHY enables us to write it in the two forms

$$\mathbf{F} = \mathbf{R}\mathbf{U} = \mathbf{V}\mathbf{R} , \tag{2.1}$$

where $\mathbf{R}$ is an orthogonal tensor, while $\mathbf{U}$ and $\mathbf{V}$ are positive-definite symmetric tensors. $\mathbf{R}$, $\mathbf{U}$, and $\mathbf{V}$ are unique. CAUCHY'S decomposition tells us that the deformation corresponding locally to $\mathbf{F}$ may be obtained by effecting pure stretches of amounts, say, v_i, along three suitable mutually orthogonal directions $\mathbf{e}_i$, followed by a rigid rotation of those directions, or by performing the same rotation first and then effecting the same stretches along the resulting directions. The quantities v_i are the principal stretches; corresponding unit proper vectors of $\mathbf{U}$ and $\mathbf{V}$ point along the principal axes of strain in the reference configuration and the present configuration χ , respectively. Indeed, if

$$\mathbf{U}\mathbf{e}_i = v_i\mathbf{e}_i , \tag{2.2}$$

then by (2.1)

$$\mathbf{V}(\mathbf{R}\mathbf{e}_i) = (\mathbf{R}\mathbf{U}\mathbf{R}^T)(\mathbf{R}\mathbf{e}_i) = v_i(\mathbf{R}\mathbf{e}_i) . \tag{2.3}$$

Thus, as just asserted, $\mathbf{U}$ and $\mathbf{V}$ have common proper numbers but different principal axes, and $\mathbf{R}$ is the rotation which carries the principal axes of $\mathbf{U}$ into the principal axes of $\mathbf{V}$. $\mathbf{R}$ is orthogonal but need not be proper orthogonal: $\mathbf{R}\mathbf{R}^T = \mathbf{1}$, so $\det \mathbf{R} = +1$ or -1, and $\det \mathbf{R}$ maintains either the one value or the other for all $\mathbf{X}$ and t, by continuity. Thus $\det \mathbf{U} = \det \mathbf{V} = |\det \mathbf{F}| = J$.

$\underset{\sim}{R}$ is called the <u>rotation tensor</u>; $\underset{\sim}{U}$ and $\underset{\sim}{V}$, the <u>right</u> and <u>left stretch tensors</u>, respectively. These tensors, like $\underset{\sim}{F}$ itself, are to be interpreted as comparing aspects of the present configuration with their counterparts in the reference configuration.

The <u>right</u> and <u>left Cauchy-Green tensors</u> $\underset{\sim}{C}$ and $\underset{\sim}{B}$ are defined as follows:

$$\underset{\sim}{C} \equiv \underset{\sim}{U}^2 = \underset{\sim}{F}^T\underset{\sim}{F} , \qquad (2.4)$$

$$\underset{\sim}{B} \equiv \underset{\sim}{V}^2 = \underset{\sim}{F}\underset{\sim}{F}^T .$$

While the fundamental decomposition (2.1) plays the major part in the proof of general theorems, calculation of $\underset{\sim}{U}$, $\underset{\sim}{V}$, and $\underset{\sim}{R}$ in special cases may be awkward, since irrational operations are usually required. $\underset{\sim}{C}$ and $\underset{\sim}{B}$, however, are calculated by mere multiplication of $\underset{\sim}{F}$ and $\underset{\sim}{F}^T$. <u>E.g.</u>, if g_{km} and $g^{\alpha\beta}$ are the covariant and contravariant metric components in arbitrarily selected co-ordinate systems in space and in the reference configuration, respectively, components of $\underset{\sim}{C}$ and $\underset{\sim}{B}$ are

$$C_{\alpha\beta} = F^k{}_\alpha F^m{}_\beta g_{km} ,$$
$$B^{km} = F^k{}_\alpha F^m{}_\beta g^{\alpha\beta}, \qquad (2.5)$$

where $F^k{}_\alpha = x^k{}_{,\alpha} \equiv \partial_{X^\alpha} \chi_k(X_1; X_2, X_3, t)$. The proper numbers of $\underset{\sim}{C}$ and $\underset{\sim}{B}$ are the squares of the principal stretches, v_i^2. The principal invariants of $\underset{\sim}{C}$ and $\underset{\sim}{B}$ are given by

$$I = \text{tr}\,\underset{\sim}{B} = \text{tr}\,\underset{\sim}{C} = v_1^2 + v_2^2 + v_3^2 ,$$

$$II = \frac{1}{2}[(\text{tr}\,\underset{\sim}{B})^2 - \text{tr}\,\underset{\sim}{B}^2] = \frac{1}{2}[\text{tr}\,\underset{\sim}{C})^2 - \text{tr}\,\underset{\sim}{C}^2] = v_1^2v_2^2 + v_2^2v_3^2 + v_3^2v_1^2 , \qquad (2.6)$$

$$III = \det \underset{\sim}{B} = \det \underset{\sim}{C} = J^2 = v_1^2v_2^2v_3^2 .$$

If we begin with the relative deformation $\underset{\sim}{F}_t$, defined by (1.39), and apply to it the polar decomposition theorem, we obtain the <u>relative rotation</u> $\underset{\sim}{R}_t$, the relative stretch tensors $\underset{\sim}{U}_t$ and $\underset{\sim}{V}_t$ and the <u>relative Cauchy-Green tensors</u> $\underset{\sim}{C}_t$ and $\underset{\sim}{B}_t$:

$$\underset{\sim}{F}_t = \underset{\sim}{R}_t \underset{\sim}{U}_t = \underset{\sim}{V}_t \underset{\sim}{R}_t \ , \quad \underset{\sim}{C}_t = \underset{\sim}{U}_t^2 \ , \quad \underset{\sim}{B}_t = \underset{\sim}{V}_t^2 \ . \tag{2.7}$$

Histories.

The restriction of the function $f(\tau)$ to times τ not later than the present time t is called the history of f up to time t and is denoted by $f^t(s)$ or f^t:

$$f^t \equiv f^t(s) \equiv f(t-s) \ , \quad t \text{ fixed} \ , \quad s \geqq 0. \tag{2.8}$$

The history f^t of f, as its name suggests, is the portion of a function of all time which corresponds to the present and past times only. Histories turn out to be of major importance in mechanics because it is the present and past that determine the future.

In the notation (2.8), for example, $\underset{\sim}{C}_t^t(s)$ is the history of the relative right Cauchy-Green tensor up to time t.

Stretching and Spin.

For a tensor defined from the relative motion, for example $\underset{\sim}{F}_t$, we introduce the notation

$$\dot{\underset{\sim}{F}}_t(t) \equiv \partial_\tau \underset{\sim}{F}_t(\tau)\Big|_{\tau=t} = -\ \partial_s \underset{\sim}{F}_t^t(s)\Big|_{s=0} \ . \tag{2.9}$$

Set

$$\begin{aligned} \underset{\sim}{G} &\equiv \dot{\underset{\sim}{F}}_t(t) \ , \\ \underset{\sim}{D} &\equiv \dot{\underset{\sim}{U}}_t(t) = \dot{\underset{\sim}{V}}_t(t) \ , \\ \underset{\sim}{W} &\equiv \dot{\underset{\sim}{R}}_t(t) \ . \end{aligned} \tag{2.10}$$

$\underset{\sim}{D}$, which is called the stretching, is the rate of change of the stretch of the configuration at time $t + \varepsilon$ with respect to that at time t, in the limit as $\varepsilon \to 0$. Likewise, $\underset{\sim}{W}$, which is called the spin, is the ultimate rate of change of the rotation from the present configuration to one occupied just before or just afterward. Since $\underset{\sim}{U}_t$ is symmetric, so is $\underset{\sim}{D}$, being its derivative with respect to a parameter:

$$\underset{\sim}{D}^T = \underset{\sim}{D} \ , \tag{2.11}$$

but, unlike $\mathbf{U}_t$, $\mathbf{D}$ generally fails to be positive-definite. If we differentiate the relation $\mathbf{R}_t(\tau)\, \mathbf{R}_t(\tau)^T = \mathbf{1}$ with respect to τ, put $\tau = t$, and use $(2.10)_4$, we find that $\mathbf{W}$ is skew:

$$\mathbf{W}^T + \mathbf{W} = \mathbf{0} . \tag{2.12}$$

From its definition $(2.10)_1$, $\mathbf{G}$ is the ultimate rate of change of $\mathbf{F}_t$, but that is not all, for by (1.40) we have

$$\mathbf{G} = \dot{\mathbf{F}}(t)\mathbf{F}(t)^{-1} . \tag{2.13}$$

Differentiation of (1.15) with respect to t yields

$$\dot{\mathbf{F}} = \nabla\dot{\chi}_{\kappa} = \nabla\dot{\mathbf{x}} = (\text{grad } \dot{\mathbf{x}})\mathbf{F} , \tag{2.14}$$

where the last step follows by the chain rule (1.27). Substitution into (2.13) yields

$$\mathbf{G} = \text{grad } \dot{\mathbf{x}} \tag{2.15}$$

We have shown that the tensor $\mathbf{G}$, defined by $(2.10)_1$, is in fact the spatial <u>velocity gradient</u>.

<u>Exercise 2.1</u>. Prove that if

$$\mathbf{G}_n \equiv \overset{(n)}{\mathbf{F}}_t(t) , \tag{2.16}$$

then

$$\mathbf{G}_n = \text{grad} \overset{(n)}{\mathbf{x}} . \tag{2.17}$$

If we differentiate the polar decomposition $(2.7)_1$ with respect to τ and then put $\tau = t$, we find that

$$\mathbf{G} = \mathbf{D} + \mathbf{W} . \tag{2.18}$$

This result, showing that $\mathbf{D}$ and $\mathbf{W}$ are the symmetric and skew parts of the velocity gradient, expresses the fundamental <u>EULER-CAUCHY-STOKES decomposition</u> of the instantaneous motion into the sum of a pure stretching along three mutually orthogonal axes and a rigid spin of those axes.

Of course, we could have defined $\mathbf{G}$ by (2.15) as the velocity gradient and $\mathbf{W}$ and $\mathbf{D}$ by (2.18) as the symmetric and skew parts of $\mathbf{G}$. We should then have had to prove $(2.10)_{2,4}$ as theorems so as to interpret $\mathbf{G}$, $\mathbf{W}$ and $\mathbf{D}$ kinematically. Writers on hydrodynamics usually prefer the argument in this order.

Motions in which $\mathbf{W} = \mathbf{0}$ are called <u>irrotational</u>. They form the main subject of study in classical hydrodynamics.

Since $\mathbf{W}$ is skew, it may be represented by an axial vector, denoted by $\mathbf{G}_\times$ or $\mathbf{W}_\times$ in GIBBS' notation and called the "vorticity" in hydrodynamics. Nowadays it seems more convenient not to introduce this vector but instead to use the tensor $\mathbf{W}$.

Further enlightenment of the difference between stretch and stretching and between rotation and spin is furnished by the following exercise.

<u>Exercise 2.2</u>. Prove that

$$\begin{aligned} \mathbf{W} &= \dot{\mathbf{R}}\mathbf{R}^T + \frac{1}{2}\mathbf{R}(\dot{\mathbf{U}}\mathbf{U}^{-1} - \mathbf{U}^{-1}\dot{\mathbf{U}})\mathbf{R}^T , \\ \mathbf{D} &= \frac{1}{2}\mathbf{R}(\dot{\mathbf{U}}\mathbf{U}^{-1} + \mathbf{U}^{-1}\dot{\mathbf{U}})\mathbf{R}^T , \end{aligned} \tag{2.19}$$

where $\mathbf{R}$ and $\mathbf{U}$ have their usual meanings as the rotation and right stretch tensors with respect to a fixed reference configuration.

Clearly the spin $\mathbf{W}$ is generally something quite different from $\dot{\mathbf{R}}$, the time rate of the rotation tensor, and the stretching $\mathbf{D}$ is entirely different from $\dot{\mathbf{U}}$, the time-rate of the stretch tensor. If the simple equations (2.19) had been available to the hydrodynamicists of the 19th century, a long and acrimonious controversy in the literature could have been avoided. HELMHOLTZ interpreted $\mathbf{W}$ as an instantaneous rotation; BERTRAND objected because a simple shear flow is rotational in HELMHOLTZ'S sense even though the particles move in straight lines, while if the particles rotate in concentric circles with an appropriate distribution of velocities, the motion is irrotational. Such controversy disappears by a glance at (2.19), which shows that the conditions $\dot{\mathbf{R}} = \mathbf{0}$ and $\mathbf{W} = \mathbf{0}$ are far from the same. The definitions (2.1) and (2.10) make the different kinematic meanings of $\mathbf{W}$ and $\dot{\mathbf{R}}$ clear and suggest that both tensors will be useful.

Higher rates of change of stretch and rotation may be defined in various ways. The most useful higher rates are the <u>Rivlin-Ericksen tensors</u> $\mathbf{A}_n$:

$$A_n \equiv \overset{(n)}{C}_t(t) , \qquad (2.20)$$

where the notation (2.9) is used.

Changes of Frame.

In classical mechanics we think of an observer as being a rigid body carrying a clock. Actually we do not need an observer as such, but only the concept of change of observer, or, as we shall say, change of frame. The ordered pair $\{x,t\}$, a place and a time, is called an event. The totality of events is space-time . A change of frame is a one-to-one mapping of space-time onto itself such that distances, time intervals, and the sense of time are preserved. We expect that every such transformation should be a time-dependent orthogonal transformation of space combined with a shift of the origin of time. This is so. The most general change of frame is given by

$$x^* = c(t) + Q(t)(x - x_o) , \quad t^* = t - a , \qquad (2.21)$$

where $c(t)$ is a time-dependent point, $Q(t)$ is a time-dependent orthogonal tensor, x_o is a fixed point, and a is a constant. We commonly say that $c(t)$ represents a change of origin, since the fixed point x_o is mapped into $c(t)$. $Q(t)$ represents a rotation and also, possibly, a reflection. Reflections are included since, although in most physics and engineering courses the student is taught to use a right-handed co-ordinate system, there is nothing in the divine order of nature to prevent two observers from orienting themselves oppositely. Of course, we may live to see the day when social democracy, in the interest of the greatest good for the greater number, forbids comrades (citizens) in eastern (western) countries to think about right-(left-) handed systems, respectively.

A frame need not be defined and certainly must not be confused with a co-ordinate system. It is convenient, however, to describe (2.21) as a change from "the unstarred frame to the starred frame", since this wording promotes the interpretation in terms of two different observers.

A quantity is said to be frame-indifferent if it is invariant under all changes of frame (2.21). More precisely, the following requirements are laid down:

$$A^* = A \quad \text{for indifferent scalars}$$
$$\mathbf{v}^* = \mathbf{Q}\mathbf{v} \quad \text{for indifferent vectors}$$
$$\mathbf{S}^* = \mathbf{Q}\mathbf{S}\mathbf{Q}^T \quad \text{for indifferent tensors (of second order)}$$

An indifferent scalar is a quantity which does not change its value. An indifferent vector is one which is the same "arrow" in the sense that

$$\text{if} \quad \mathbf{v} = \mathbf{x} - \mathbf{y} , \quad \text{then} \quad \mathbf{v}^* = \mathbf{x}^* - \mathbf{y}^* . \tag{2.22}$$

By (2.21), then,

$$\mathbf{v}^* = \mathbf{Q}(\mathbf{x} - \mathbf{y}) = \mathbf{Q}\mathbf{v} , \tag{2.23}$$

as asserted. An indifferent tensor is one that transforms indifferent vectors into indifferent vectors. That is,

$$\text{if} \quad \mathbf{v} = \mathbf{S}\mathbf{w} \quad \text{and} \quad \mathbf{v}^* = \mathbf{Q}\mathbf{v} , \quad \mathbf{w}^* = \mathbf{Q}\mathbf{w} , \quad \text{then} \quad \mathbf{v}^* = \mathbf{S}^*\mathbf{w}^* . \tag{2.24}$$

By substituting the first three equations into the last, we find that

$$\mathbf{Q}\mathbf{v} = \mathbf{S}^*\mathbf{Q}\mathbf{w} = \mathbf{Q}\mathbf{S}\mathbf{w} . \tag{2.25}$$

Since this relation is to hold for all $\mathbf{w}$, we infer the rule $\mathbf{S}^*\mathbf{Q} = \mathbf{Q}\mathbf{S}$, as stated above.

In mechanics we meet some quantities that are indifferent and some that are not. Sometimes we have a vector or tensor defined in one frame only. By using the above rules, we may extend the definition to all frames in such a way as to obtain a frame-indifferent quantity. Such an extension is trivial. Usually, however, we are given a definition valid in all frames from the start. In that case, we have to find out what transformation law is obeyed and thus determine whether or not the quantity is frame-indifferent.

Consider, for example, the motion of a body. Under the change of frame (2.21), (1.1) becomes

$$\mathbf{x}^* = \mathbf{c}(t) + \mathbf{Q}(t)[\boldsymbol{\chi}(X,t) - x_o] \equiv \boldsymbol{\chi}^*(X,t^*) . \tag{2.26}$$

Differentiation with respect to t^* yields

$$\dot{\mathbf{x}}^* = \dot{\mathbf{c}} + \dot{\mathbf{Q}}[\chi - \mathbf{x}_o] + \mathbf{Q}\dot{\mathbf{x}} , \tag{2.27}$$

so that, by (2.26)

$$\dot{\mathbf{x}}^* - \mathbf{Q}\dot{\mathbf{x}} = \dot{\mathbf{c}} + \mathbf{A}(\mathbf{x}^* - \mathbf{c}) , \tag{2.28}$$

where

$$\mathbf{A} = \dot{\mathbf{Q}}\mathbf{Q}^T = -\mathbf{A}^T . \tag{2.29}$$

$\mathbf{A}$ is called the <u>angular velocity</u> or <u>spin</u> of the starred frame with respect to the unstarred one. The result (2.28) shows that the velocity is not a frame-indifferent quantity, since if it were, the right-hand side of (2.28) would have to be replaced by $\mathbf{0}$. Likewise, the acceleration is not frame-indifferent.

<u>Exercise 2.3</u>. Prove that

$$\ddot{\mathbf{x}}^* - \mathbf{Q}\ddot{\mathbf{x}} = \ddot{\mathbf{c}} + 2\mathbf{A}(\dot{\mathbf{x}}^* - \dot{\mathbf{c}}) + (\dot{\mathbf{A}} - \mathbf{A}^2)(\mathbf{x}^* - \mathbf{c}) . \tag{2.30}$$

In (2.26) we may refer both the motions χ and χ^* to the same reference configuration κ if we wish to. This amounts to replacing χ and χ^* by χ_κ and $\chi_\kappa{}^*$, respectively, since these functions have the same values as the former. By taking the gradient of the resulting formula we obtain

$$\mathbf{F}^* = \mathbf{QF} . \tag{2.31}$$

From the polar decomposition (2.1), then ,

$$\mathbf{R}^*\mathbf{U}^* = \mathbf{QRU} . \tag{2.32}$$

Since $\mathbf{QR}$ is orthogonal and since a polar decomposition is unique,

$$\mathbf{R}^* = \mathbf{QR} , \quad \text{and} \quad \mathbf{U}^* = \mathbf{U} . \tag{2.33}$$

Hence

$$\begin{aligned} \mathbf{V}^* = \mathbf{R}^*\mathbf{U}^*\mathbf{R}^{*T} &= \mathbf{QRU}(\mathbf{QR})^T , \\ &= \mathbf{QVQ}^T . \end{aligned} \tag{2.34}$$

Thus we have shown that $\mathbf{V}$ is frame-indifferent, while $\mathbf{F}$, $\mathbf{R}$, and $\mathbf{U}$ are not. Of course, $\mathbf{C}^* = \mathbf{C}$ and $\mathbf{B}^* = \mathbf{QBQ}^T$, as is immediate by applying $(2.33)_2$ and (2.34) to the definitions (2.4).

If we differentiate (2.31) with respect to time, we find that

$$\dot{\mathbf{F}}^* = \mathbf{Q}\dot{\mathbf{F}} + \dot{\mathbf{Q}}\mathbf{F} , \tag{2.35}$$

but by (2.13) $\dot{\mathbf{F}} = \mathbf{GF}$, so

$$G^*F^* = QGF + \dot{Q}F ,$$
$$= QGQ^T F^* + \dot{Q}Q^T F^* \quad . \qquad (2.36)$$

Since F^* is non-singular, it may be cancelled from this equation, which by use of (2.18) becomes

$$D^* + W^* = Q(D + W)Q^T + A , \qquad (2.37)$$

where A is the spin (2.29) of the starred frame with respect to the unstarred one. Since a decomposition into symmetric and skew parts is unique,

$$D^* = QDQ^T , \quad W^* = QWQ^T + A . \qquad (2.38)$$

These formulae embody the theorem of ZORAWSKI: The stretching is frame-indifferent, while the spin in the starred frame is the sum of the spin in the unstarred frame and the spin of the starred frame with respect to the unstarred. The assertion is intuitively plain, since a change of frame is a rigid motion, which does not alter the stretchings of elements though it does rotate the directions in which they seem to occur.

Exercise 2.4. Prove that the Rivlin-Ericksen tensors A_n are frame-indifferent.

In the next lecture most relations of interest to us shall be invariant under change of frame. To have to demand invariance of a relation under a group of transformations is a sign that we have not formulated a good mathematical language for the problem at hand, since in such a language invariance is automatic. The use of a frame in classical mechanics is artificial in that the frame and its motion really have nothing to do with the phenomenon being observed. An abstract formulation of space-time, kinematics, and dynamics, in which frames play no part, has been given by NOLL in his recent course at Bressanone. In these lectures, however, I shall continue to use a formulation of mechanics close to the one taught to freshmen before the brilliant young nuclear physicists began their program of revision by excision.

LECTURE 3: FORCES, CONSTITUTIVE EQUATIONS, SIMPLE MATERIALS.

Forces and Moments.

Of all the axioms of mechanics, those for systems of forces and torques are the most difficult to formulate. Forces are vectors of a special kind. The usual treatments of mechanics do not even mention explicitly the difference between forces and more general vectors and do not write out any formal properties of forces. The treatment I present now is likewise an informal one, though a little more careful.

Forces and torques are among the primitive elements of mechanics; like bodies and motions, they are given a priori. While several kinds of forces and torques are considered in a more general mechanics, in these lectures we shall need only the classical ones. Forces act on the parts $\mathcal{P}$ of a body $\mathcal{B}$ in a configuration χ. We shall require two kinds of forces: body force $f_b(\mathcal{P})$, which is an absolutely continuous function of the volume of $\mathcal{P}_\chi$, and contact force $f_c(\mathcal{P})$, which is an absolutely continuous function of the surface area of the boundary $\partial\mathcal{P}_\chi$ of $\mathcal{P}_\chi$. The resultant force $f(\mathcal{P})$ acting on $\mathcal{P}$ in χ is given by

$$f(\mathcal{P}) = f_b(\mathcal{P}) + f_c(\mathcal{P}), \tag{3.1}$$

where

$$f_b(\mathcal{P}) = \int_{\mathcal{P}_\chi} b\,dm = \int_{\mathcal{P}_\chi} \rho b\,dv, \tag{3.2}$$

$$f_c(\mathcal{P}) = \int_{\partial\mathcal{P}_\chi} t\,ds.$$

The two densities, $\underset{\sim}{b}$ and $\underset{\sim}{t}$, are called the <u>specific body force</u> and the <u>traction</u>, respectively.

The <u>resultant moment of force</u> $\underset{\sim}{L}(\mathcal{P};\underset{\sim}{x}_0)$ with respect to $\underset{\sim}{x}_0$ is defined by

$$\underset{\sim}{L}(\mathcal{P},\underset{\sim}{x}_0) = \int_{\mathcal{P}_{\underset{\sim}{\chi}}} (\underset{\sim}{x} - \underset{\sim}{x}_0) \wedge \underset{\sim}{b}dm + \int_{\partial\mathcal{P}_{\underset{\sim}{\chi}}} (\underset{\sim}{x} - \underset{\sim}{x}_0) \wedge \underset{\sim}{t}ds. \tag{3.3}$$

A moment or <u>simple torque</u> is only a special case of a <u>torque</u>, but more general torques are not needed in this course.

<u>Exercise 3.1</u>. Prove that

$$\underset{\sim}{L}(\mathcal{P},\underset{\sim}{x}_0') = \underset{\sim}{L}(\mathcal{P},\underset{\sim}{x}_0) + (\underset{\sim}{x}_0 - \underset{\sim}{x}_0') \wedge \underset{\sim}{f}(\mathcal{P}) ; \tag{3.4}$$

We regard forces and moments of force as quantities given <u>a priori</u> and independent of the observer. Thus we assume that <u>forces and torques are indifferent</u>. That is, under a change of frame (2.21),

$$\underset{\sim}{b}^* = \underset{\sim}{Q}\underset{\sim}{b} \qquad \text{and} \qquad \underset{\sim}{t}^* = \underset{\sim}{Q}\underset{\sim}{t}. \tag{3.5}$$

<u>Euler's Laws of Mechanics</u>.

The <u>momentum</u> $\underset{\sim}{m}(\mathcal{P})$ and the <u>moment of momentum</u> $\underset{\sim}{M}(\mathcal{P};\underset{\sim}{x}_0)$ of $\mathcal{P}$ in the configuration $\underset{\sim}{\chi}(\mathcal{B},t)$ are defined by

$$\begin{aligned} \underset{\sim}{m}(\mathcal{P}) &= \int_{\mathcal{P}_{\underset{\sim}{\chi}}} \dot{\underset{\sim}{x}}dm, \\ \underset{\sim}{M}(\mathcal{P};\underset{\sim}{x}_0) &= \int_{\mathcal{P}_{\underset{\sim}{\chi}}} (\underset{\sim}{x} - \underset{\sim}{x}_0) \wedge \dot{\underset{\sim}{x}}dm. \end{aligned} \tag{3.6}$$

As axioms relating the forces applied to the motion produced we lay down <u>EULER'S laws</u>:

$$\begin{aligned} \underset{\sim}{f}(\mathcal{P}) &= \dot{\underset{\sim}{m}}(\mathcal{P}) \\ \underset{\sim}{L}(\mathcal{P};\underset{\sim}{x}_0) &= \dot{\underset{\sim}{M}}(\mathcal{P};\underset{\sim}{x}_0). \end{aligned} \tag{3.7}$$

(Of course, these laws generalize and include much earlier ones, due in varying forms and circumstances to mediaeval schoolmen, HUYGENS, NEWTON, JAMES BERNOULLI, and others. There is no justice in emphasizing our debt to any one of these predecessors of EULER to the exclusion of the rest.) EULER's laws assert that the resultant force equals the rate of change of momentum, and the resultant moment of force equals the rate of change of moment of momentum. Here $\mathcal{P}$ is a fixed part of the body $\mathcal{B}$, and $\mathbf{x}_0$ is a fixed point in space. The dots on the right-hand side of (3.7) indicate the ordinary time derivatives. (Of course $\mathcal{P}_{\chi}$ is not a fixed region in space, since it is the configuration of $\mathcal{P}$ at time t in the motion χ.)

Exercise 3.2. If $(3.7)_1$ holds, then $(3.7)_2$ holds for one $\mathbf{x}_0$ if and only if it holds for all $\mathbf{x}_0$.

It is possible to derive EULER's laws as a theorem from a single axiom proposed by NOLL: The rate of working is frame-indifferent for every part $\mathcal{P}$ of every body $\mathcal{B}$.

In modern mechanics the definitions of $\mathbf{m}$ and $\mathbf{M}$ are sometimes generalized, and torques more general than (3.3) are often considered. EULER's laws, or an equivalent statement, remain the fundamental axioms for all kinds of mechanics. In particular, if the sign $\int$ is taken to mean a Lebesgue-Stieltjes integral, EULER's laws may be shown to imply as corollaries the so-called "Newtonian equations" of analytical dynamics, provided certain assumptions be made about the forces.

The Euler-Cauchy Stress Principle.

As defined, the densities $\mathbf{b}$ and $\mathbf{t}$ in (3.2) may be extremely general:

$$\mathbf{b} = \mathbf{b}(\mathbf{x},t,\mathcal{P},\mathcal{B}), \qquad \mathbf{t} = \mathbf{t}(\mathbf{x},t,\mathcal{P},\mathcal{B}). \tag{3.8}$$

We shall restrict attention to body force densities that are unaffected by the presence or absence of bodies in space:

$$\mathbf{b} = \mathbf{b}(\mathbf{x},t), \tag{3.9}$$

whatever be $\mathcal{P}$ and $\mathcal{B}$. Such body forces are called external. (A particular kind of more general body force called mutual, such as universal gravitation, is sometimes included in mechanics but will not be needed in this course.) The particular case when $\mathbf{b}$ = constant pertains to heavy bodies. If $\mathbf{b}$ is the gradient of a scalar field, the body force is said to be conservative. We shall restrict attention also to a particular kind of contact force, namely, one such that the traction $\mathbf{t}$ at any given place and time has a common value for all parts p having a common tangent plane and lying upon the same side of it:

$$\mathbf{t} = \mathbf{t}(\mathbf{x},t,\mathbf{n}), \tag{3.10}$$

where $\mathbf{n}$ is the outer normal to ∂p in the configuration χ. Such tractions are called simple. The pioneers of continuum mechanics laboriously created the assumption (3.10) through abstraction from special cases. To them it seemed ultimate in generality. The assumption embodied in (3.1), $(3.2)_3$, and (3.10) is the EULER-CAUCHY stress principle, which is the keystone of classical continuum mechanics. We shall not need to depart from it in this course.

The stress principle is put to use through CAUCHY's fundamental lemma: There exists a tensor $\mathbf{T}(\mathbf{x},t)$, called the stress tensor, such that

$$\mathbf{t}(\mathbf{x},t,\mathbf{n}) = \mathbf{T}(\mathbf{x},t)\mathbf{n}. \tag{3.11}$$

That is, the traction $\mathbf{t}$, which at the outset was allowed to depend arbitrarily on the normal $\mathbf{n}$, is in fact a linear homogeneous function of it.

Exercise 3.3. Show that CAUCHY's fundamental lemma follows from (3.10) and either $(3.7)_1$ or $(3.7)_2$.

Cauchy's Laws of Continuum Mechanics.

As a result of CAUCHY's fundamental lemma, for continua subject to simple tractions, simple torques, and external body forces we may write EULER's laws (3.7) in the following special and significant forms:

$$\left(\int_p \dot{\mathbf{x}}\,dm\right)^{\cdot} = \int_{\partial p_\chi} \mathbf{T}\mathbf{n}\,ds + \int_{p_\chi} \mathbf{b}\,dm,$$

$$[\int_{\mathcal{P}} (\underset{\sim}{x} - \underset{\sim}{x}_0) \wedge \dot{\underset{\sim}{x}}dm]^{\cdot} = \int_{\partial \mathcal{P}_{\underset{\sim}{\chi}}} (\underset{\sim}{x} - \underset{\sim}{x}_0) \wedge \underset{\sim}{T}\underset{\sim}{n}ds + \int_{\mathcal{P}_{\underset{\sim}{\chi}}} (\underset{\sim}{x} - \underset{\sim}{x}_0) \wedge \underset{\sim}{b}dm.$$

for every part $\mathcal{P}$ of every body $\mathcal{B}$. The subscript $\underset{\sim}{\chi}$ is omitted from $\mathcal{P}$ on the left-hand side as a reminder that the time derivative is calculated for a fixed part $\mathcal{P}$ of the body, not for a fixed region of space.

These equations are both of the form called an <u>equation of balance</u> for a tensor field Ψ:

$$(\int_{\mathcal{P}} \Psi dm)^{\cdot} = \int_{\partial \mathcal{P}_{\underset{\sim}{\chi}}} \underset{\sim}{E}\{\Psi\}\underset{\sim}{n}ds + \int_{\mathcal{P}_{\underset{\sim}{\chi}}} s\{\Psi\}dm. \tag{3.13}$$

$\underset{\sim}{E}\{\Psi\}$, a tensor of order 1 greater than Ψ, is called an <u>efflux</u> of Ψ, while $s\{\Psi\}$ is called a <u>source</u> of Ψ. An equation of balance expresses the rate of growth of $\int_{\mathcal{P}} \Psi dm$ as the sum of two parts: a rate of flow $-\underset{\sim}{E}\{\Psi\}$ inward through the boundary $\partial \mathcal{P}_{\chi}$ of the configuration $\mathcal{P}_{\chi}$ of $\mathcal{P}$ and a creation $s\{\Psi\}$ in the interior of that configuration. Equations of this form occur frequently in mathematical physics. Under the assumptions of smoothness made at the beginning of this course, an equation of balance is always equivalent to a differential equation.

<u>Exercise 3.4</u>. Prove that

$$(\int_{\mathcal{P}} \Psi dm)^{\cdot} = \int_{\mathcal{P}_{\underset{\sim}{\chi}}} \dot{\Psi} dm, \tag{3.14}$$

where $\dot{\Psi}$ is the derivative of Ψ in the material description and hence may be calculated from (1.25). Show that the equation of balance (3.13) holds for every part $\mathcal{P}$ of every body $\mathcal{B}$ if and only if at each interior point of $\mathcal{B}$

$$\rho \dot{\Psi} = \operatorname{div} \underset{\sim}{E}\{\Psi\} + \rho s\{\Psi\}. \tag{3.15}$$

The general differential equation (3.15) is a consequence of the divergence theorem, an easy one. Textbooks for engineers and physicists still seem to prefer proofs of each special application by means of diagrams showing boxes decorated by many small arrows. Such proofs are regarded as "intuitive", apparently because they include, over and over again, a bad proof of the divergence theorem itself along with the particular application of it. "Intuition" of this kind must be bought at the price of more time and boredom than a course on modern mechanics can afford. We must remark that (3.15) holds subject to specific assumptions. Use of the divergence theorem requires assumptions about the region and about the fields. The assumptions about the region are satisfied here, because only interior points are considered; in fact, it would suffice to consider only those parts p whose configurations are spheres about the point in question. The fields, however, must be smooth. It suffices to assume that $\underset{\sim}{E}$ is continuously differentiable and that $\rho\dot{\Psi}$ and ρs are continuous in a sufficiently small sphere about the interior point considered. In general, (3.15) does not hold at points of $\partial\mathcal{B}_{\underset{\sim}{\chi}}$ or at interior points where any of the fields ρ, $\dot{\Psi}$, $\underset{\sim}{E}$, s fail to exist or to be sufficiently smooth.

If we apply (3.15) to $(3.12)_1$, we conclude at once that EULER's first law holds if and only if

$$\rho\ddot{\underset{\sim}{x}} = \operatorname{div}\underset{\sim}{T} + \rho\underset{\sim}{b} \tag{3.16}$$

at interior points of $\mathcal{B}_{\underset{\sim}{\chi}}$. This equation expresses <u>CAUCHY's first law of continuum mechanics</u>. To apply (3.15) to $(3.12)_2$, we let $\underset{\sim}{M}$ stand for the tensor such that $(\underset{\sim}{x} - \underset{\sim}{x}_0) \wedge (\underset{\sim}{T}\underset{\sim}{n}) = \underset{\sim}{M}\underset{\sim}{n}$ and conclude that EULER's second law holds if and only if

$$\begin{aligned}\rho(\underset{\sim}{x} - \underset{\sim}{x}_0) \wedge \ddot{\underset{\sim}{x}} &= \operatorname{div}\underset{\sim}{M} + \rho(\underset{\sim}{x} - \underset{\sim}{x}_0) \wedge \underset{\sim}{b},\\ &= \underset{\sim}{T}^T - \underset{\sim}{T} + (\underset{\sim}{x} - \underset{\sim}{x}_0) \wedge \operatorname{div}\underset{\sim}{T} + \rho(\underset{\sim}{x} - \underset{\sim}{x}_0) \wedge \underset{\sim}{b},\end{aligned} \tag{3.17}$$

where the second form follows by an easy identity. If (3.16) holds, then

$$\underset{\sim}{T}^T = \underset{\sim}{T} : \tag{3.18}$$

The stress tensor is symmetric, and conversely. This equation expresses <u>CAUCHY's second law of continuum mechanics</u>.

Since this second law, in particular, has been questioned from time to time, I pause to emphasize its special character. We began with the general laws of EULER, but we applied them only subject to some special assumptions:

1. All torques are moments of forces.
2. The traction is simple.

(We assumed also that the body force is external, but that restriction is not important here.) Under these assumptions, EULER's laws are equivalent to CAUCHY's laws when χ, $\mathbf{b}$, $\rho\ddot{\mathbf{x}}$, and $\mathbf{T}$ are sufficiently smooth. CAUCHY's first law expresses locally the balance of linear momentum. CAUCHY's second law, if the first is satisfied, expresses locally the balance of moment of momentum.

Equivalent Processes.

We now consider motions and forces which obey the laws of mechanics. Formally, the motion of a body and the forces acting upon the corresponding configurations of the body constitute a dynamical process if CAUCHY's laws (3.16) and (3.18) are satisfied. If the body and its mass distribution are given, CAUCHY's first law (3.16) determines a unique body force $\mathbf{b}$ for any particular stress and motion. While in some specific case $\mathbf{b}$ will be given, here we wish to consider rather the totality of all possible problems. There is then no reason to restrict $\mathbf{b}$. Consequently, any pair of functions $\{\chi, \mathbf{T}\}$, where χ is a mapping of the body $\mathcal{B}$ onto configurations in space and where $\mathbf{T}$ is any smooth field of symmetric tensors defined at each time t over the configuration $\mathcal{B}_\chi$, defines a dynamical process.

Under a change of frame, χ becomes χ^* as given by (2.26). Our assumption (3.5) asserts that body force and contact force are indifferent. Since, of course, the normal $\mathbf{n}$ is indifferent, Cauchy's lemma (3.11) implies that the stress tensor $\mathbf{T}$ is indifferent, so that by a result in Lecture 2, $\mathbf{T} = \mathbf{QTQ}^T$. Two dynamical processes $\{\chi, \mathbf{T}\}$ and $\{\chi^*, \mathbf{T}^*\}$ related in this way are regarded as the same motion and associated contact forces as seen by two different observers. Thus, formally, two dynamical processes $\{\chi, \mathbf{T}\}$ and $\{\chi^*, \mathbf{T}^*\}$ will be called equivalent if they are related as follows:

$$\chi^*(X,t^*) = c(t) + Q(t)[\chi(X,t) - x_0],$$
$$t^* = t - a, \tag{3.19}$$
$$T^*(X,t^*) = Q(t)T(X,t)Q(t)^T,$$

where the notations are those introduced in Lecture 2. We here regard the stress tensor as a function of the particle **X** and the time t.

Constitutive Equations.

As I said in the first lecture, the general principles of mechanics apply to all bodies and motions, and the diversity of materials in nature is represented in the theory by constitutive equations. A constitutive equation is a relation between forces and motions. In popular terms, forces applied to a body "cause" it to undergo a motion, and the motion "caused" differs according to the nature of the body. In continuum mechanics the forces of interest are contact forces, which are specified by the stress tensor T. Just as different figures are defined in geometry as idealizations of certain important natural objects, in continuum mechanics ideal materials are defined by particular relations between the stress tensor and the motion of the body. Some materials are important in themselves, but most of them are of more interest as members of a class than in detail. Thus a general theory of constitutive equations is needed. I shall now present certain aspects of NOLL's theory, which was first published in 1958. His axioms have some prior history, as is only to be expected, since the need for a general method is never felt until successes have been achieved in special cases, which then call for unification. A good general principle is always applied before it is stated.

Axioms for Constitutive Equations.

1. Principle of determinism. The stress at the particle X in the body $\mathcal{B}$ at time t is determined by the history χ^t of the motion of $\mathcal{B}$ up to time t:

$$T(X,t) = \mathfrak{F}(\chi^t; X,t). \tag{3.20}$$

Here $\mathfrak{F}$ is a functional in the most general sense of the term, namely, a rule of correspondence. (3.20) asserts that the motion of the body up to and including the present time determines a unique symmetric stress tensor $\underset{\sim}{T}$ at each point of the body, and the manner in which it does so may depend upon X and t. The functional $\mathfrak{F}$ is called the <u>constitutive functional</u>, and (3.20) is the <u>constitutive equation</u> of the ideal material defined by $\mathfrak{F}$. Notice that the past, as much of it as need be, may affect the present stress, but in general past and future are not interchangeable. The common prejudice that mechanics concerns phenomena reversible in time is too naive to need refuting.

2. <u>Principle of local action</u>. In the principle of determinism, the motions of particles Z which lie far away from X are allowed to affect the stress at X. The notion of contact force makes it natural to exclude action at a distance as a material property. Accordingly, we assume as second axiom of continuum mechanics that the motion of particles at a finite distance from X in some configuration may be disregarded in calculating the stress at X. (Of course, by the smoothness assumed for $\underset{\sim}{\chi}$, particles once a finite distance apart are always a finite distance apart.) Formally, if

$$\bar{\underset{\sim}{\chi}}^t(Z,s) = \underset{\sim}{\chi}^t(Z,s) \text{ when } s \geqq 0 \text{ and } Z \varepsilon N(X), \tag{3.21}$$

where $N(X)$ is some neighborhood of X, then

$$\mathfrak{F}(\bar{\underset{\sim}{\chi}}^t; X,t) = \mathfrak{F}(\underset{\sim}{\chi}^t; X,t). \tag{3.22}$$

3. <u>Principle of material frame-indifference</u>. We have said that we shall regard two equivalent dynamical processes as being really the same process, viewed by two different observers. We regard material properties as being likewise indifferent to the choice of observer. Since constitutive equations are designed to express idealized material properties, we require that they shall be frame-indifferent. That is, if (3.20) holds, <u>viz</u>

$$\underset{\sim}{T}(X,t) = \mathfrak{F}(\underset{\sim}{\chi}^t; X,t), \tag{3.20}$$

then the constitutive functional $\mathfrak{J}$ must be such that

$$\mathbf{T}(X,t^*)^* = \mathfrak{J}[(\boldsymbol{\chi}^{t^*})^*; X,t^*), \qquad (3.23)$$

where $\{\chi^*, T^*\}$ is any dynamical process equivalent to $\{\boldsymbol{\chi}, \mathbf{T}\}$. Like the principle of local action, the principle of frame-indifference imposes a restriction on the functionals $\mathfrak{J}$ to be admitted in constitutive equations. Namely, $\mathfrak{J}$ must be such that the constitutive equation (3.20) is invariant under the transformations (3.19). Only those functionals $\mathfrak{J}$ that satisfy the requirements of local action and frame-indifference are admissible as constitutive functionals.

Some steps may be taken to delimit the class of functionals satisfying these axioms, but in these lectures I shall treat only a special case, which is still general enough to include all the older theories of continua and most of the more recent ones. The special case is called the simple material.

Simple Materials.

A motion $\boldsymbol{\chi}$ is homogeneous with respect to the reference configuration $\boldsymbol{\kappa}$ if

$$\mathbf{x} = \boldsymbol{\chi}_{\boldsymbol{\kappa}}(\mathbf{X},t) = \mathbf{F}(t)(\mathbf{X} - \mathbf{X}_0) + \mathbf{x}_0(t), \qquad (3.24)$$

where $\mathbf{x}_0(t)$ is a place in $\boldsymbol{\chi}$, possibly moving, $\mathbf{X}_0$ is a fixed place in the reference configuration $\boldsymbol{\kappa}$, and $\mathbf{F}(t)$, which is the deformation gradient, is a non-singular tensor that does not depend on $\mathbf{X}$. A motion is homogeneous if and only if it carries every straight line at time 0 into a straight line at time t. A motion homogeneous with respect to one reference configuration generally fails to be so with respect to another.

Physically minded people almost always assume that everything there is to know about a material can be found out by performing experiments on homogeneous motions of a body of that material from whatever state they happen to find it in. The materials in the special class that conforms to their prejudices are called simple materials. Formally, the material defined by (3.20) is called simple if there exists a reference configuration $\boldsymbol{\kappa}$ such that

$$\mathbf{T}(X,t) = \mathfrak{F}(\chi^t;\ X,t) = \underset{s=0}{\overset{\infty}{\mathfrak{G}_{\kappa}}}[\mathbf{F}^t(X,s);X]. \tag{3.25}$$

That is, the stress at the place $\mathbf{x}$ occupied by the particle X at time t is determined by the totality of deformation gradients with respect to κ experienced by that particle up to the present time. $\mathfrak{G}_{\kappa}$ is called a <u>response functional</u> of the simple material. Ordinarily we shall write (3.25) in one of the simpler forms

$$\mathbf{T} = \underset{s=0}{\overset{\infty}{\mathfrak{G}_{\kappa}}}[\mathbf{F}^t(\delta)] = \mathfrak{G}_{\kappa}(\mathbf{F}^t) = \mathfrak{G}(\mathbf{F}^t), \tag{3.26}$$

with exactly the same meaning.

The defining equation (3.26) asserts that at a given particle, $\mathfrak{F}$ depends on the motion, χ^t, only through its deformation gradient, $\mathbf{F}^t$, with respect to some fixed reference configuration. That is, all motions having the same gradient history at a given particle and time give rise to the same stress at that particle and time. Since, trivially, a homogeneous motion with any desired gradient history may be constructed, only homogeneous motions need be considered in determining any material properties that the original constitutive functional $\mathfrak{F}$ may describe.

We remark first that it is unnecessary to mention a particular reference configuration in the above definition. In Lecture 1 we obtained the important formula

$$\mathbf{F}_1 = \mathbf{F}_2\mathbf{P} \tag{1.32}$$

connecting the gradients with respect to two reference configurations κ_1 and κ_2. By (3.26), then,

$$\mathbf{T} = \mathfrak{G}_{\kappa_1}(\mathbf{F}_1^t) = \mathfrak{G}_{\kappa_1}(\mathbf{F}_2^t\mathbf{P}). \tag{3.27}$$

Since $\mathbf{P}$ is the gradient of the transformation from κ_1 to κ_2, it is constant in time, and the right-hand side of (3.27) gives $\mathbf{T}$ as a functional of the history $\mathbf{F}_2^t$ of the deformation gradient with respect to κ_2. Thus if we write κ in (3.26) as κ_1, and if we set

$$\mathfrak{G}_{\kappa_2}(F^t) = \mathfrak{G}_{\kappa_1}(F^t P), \tag{3.28}$$

we see that a relation of the form (3.26) holds again if we take κ_2 as reference. Therefore, we may speak of a simple material without mentioning any particular reference configuration, and usually we do not write the subscript κ on $\mathfrak{G}$. We must recall, however, that for a given simple material with constitutive functional $\mathfrak{J}$ there are infinitely many different response functionals $\mathfrak{G}_\kappa$, one for each choice of reference configuration κ.

Next we remark that, trivially, the simple material satisfies the principles of determinism and local action, no matter what be the response functional $\mathfrak{G}$. It is not so for the principle of material frame-indifference, as we shall see in the next lecture.

The theory of simple materials includes all the common theories of continua studied in works on engineering, physics, applied mathematics, *etc*. *E.g.*, the *elastic material* is defined by the special case when the functional $\mathfrak{G}$ reduces to a function $\mathfrak{g}$ of the present deformation gradient $F(X,t)$:

$$T = \mathfrak{g}(F,X), \tag{3.29}$$

where $\mathfrak{g}$ is a function. The *linearly viscous material* is defined by the slightly more general case when $\mathfrak{G}$ reduces to a function of $F(t)$ and $\dot{F}(t)$ which is linear in $\dot{F}$:

$$T = K(F,X)[\dot{F}] = L(F,X)[G], \tag{3.30}$$

where the second form follows from the first by (2.13). The Boltzmann accumulative theory is obtained by supposing that $\mathfrak{G}$ is expressible as an integral. It is customary to restrict these theories still further by assuming that $|F - 1|$ is small or imposing requirements of material symmetry, or both, as we shall see later. Many but not all the recent non-linear theories are included as special cases in the theory of simple materials. The simple material represents, in general, a material with long-range memory, so that stress relaxation, creep, and fatigue can occur.

Exercise 3.5. Prove that (3.30) satisfies the principle of material frame-indifference if and only if it is equivalent to

$$\underset{\sim}{R}^T \underset{\sim}{T}\underset{\sim}{R} = L(\underset{\sim}{C},\underset{\sim}{X})[\underset{\sim}{R}^T \underset{\sim}{D}\underset{\sim}{R}], \tag{3.31}$$

where L is a linear operator.

Our interest here, however, is less on the physical side than the mathematical. There are plenty of persons who, faced by any theory whatever, will easily find and talk about cases and situations it is not general enough to include. A proved theorem outweighs an infinity of physical talk. It is in the theory of simple materials that the main mathematical developments have occurred, and the remaining lectures will consider them exclusively.

LECTURE 4: REDUCED CONSTITUTIVE EQUATIONS. INTERNAL CONSTRAINTS.

Reduction for Material Frame-Indifference.

Under a change of frame, $\underset{\sim}{F}^* = \underset{\sim}{Q}\underset{\sim}{F}$, as we have seen in Lecture 2. Let $\underset{\sim}{Q}^t(s)$ be the history up to time t of any orthogonal tensor function $\underset{\sim}{Q}(t)$. Then the principle of material frame-indifference states that the response functional $\underset{\sim}{\mathfrak{G}}$ of a simple material must satisfy the equation

$$\underset{s=o}{\overset{\infty}{\underset{\sim}{\mathfrak{G}}}}[\underset{\sim}{Q}^t(s)\underset{\sim}{F}^t(s)] = \underset{\sim}{Q}(t)\underset{s=o}{\overset{\infty}{\underset{\sim}{\mathfrak{G}}}}[\underset{\sim}{F}^t(s)]\ \underset{\sim}{Q}(t)^T , \tag{4.1}$$

identically in the orthogonal tensor history $\underset{\sim}{Q}^t(s)$ and the non-singular tensor history $\underset{\sim}{F}^t(s)$.

We can solve this equation. According to the polar decomposition theorem, $\underset{\sim}{F}^t(s) = \underset{\sim}{R}^t(s)\ \underset{\sim}{U}^t(s)$, so

$$\underset{s=o}{\overset{\infty}{\underset{\sim}{\mathfrak{G}}}}[\underset{\sim}{F}^t(s)] = \underset{\sim}{Q}(t)^T\underset{s=o}{\overset{\infty}{\underset{\sim}{\mathfrak{G}}}}[\underset{\sim}{Q}^t(s)\underset{\sim}{R}^t(s)\underset{\sim}{U}^t(s)]\ \underset{\sim}{Q}(t) . \tag{4.2}$$

Since this equation must hold for all $\underset{\sim}{Q}^t$, $\underset{\sim}{R}^t$, $\underset{\sim}{U}^t$, it must hold in particular if $\underset{\sim}{Q}^t = (\underset{\sim}{R}^t)^T$. Therefore

$$\underset{\sim}{\mathfrak{G}}(\underset{\sim}{F}^t) = \underset{\sim}{R}(t)\underset{\sim}{\mathfrak{G}}(\underset{\sim}{U}^t)\underset{\sim}{R}(t)^T . \tag{4.3}$$

Conversely, suppose $\underset{\sim}{\mathfrak{G}}$ is of this form, and consider an arbitrary orthogonal tensor history $\underset{\sim}{Q}^t$. Since the polar decomposition of $\underset{\sim}{Q}^t\underset{\sim}{F}^t$ is $(\underset{\sim}{Q}^t\underset{\sim}{R}^t)\underset{\sim}{U}^t$,

$$\begin{aligned}\underset{\sim}{\mathfrak{G}}(\underset{\sim}{Q}^t\underset{\sim}{F}^t) &= \underset{\sim}{Q}(t)\underset{\sim}{R}(t)\underset{\sim}{\mathfrak{G}}(\underset{\sim}{U}^t)[\underset{\sim}{Q}(t)\underset{\sim}{R}(t)]^T ,\\ &= \underset{\sim}{Q}(t)\underset{\sim}{\mathfrak{G}}(\underset{\sim}{F}^t)\underset{\sim}{Q}(t)^T ,\end{aligned} \tag{4.4}$$

so that (4.1) is satisfied. Therefore, (4.3) gives the general solution of the functional equation (4.1). We have proved then, that <u>the constitutive equation of a simple material may be put into the form</u>

$$\underset{\sim}{T} = \underset{\sim}{R}\underset{\sim}{\mathfrak{G}}(\underset{\sim}{U}^t)\underset{\sim}{R}^T , \tag{4.5}$$

and, conversely, that any functional $\mathfrak{G}$ of positive-definite symmetric tensor histories, if its values are symmetric tensors, serves to define a simple material through (4.5). A constitutive equation of this kind, in which the functionals or functions occurring are not subject to any further restriction, is called a reduced form.

The result (4.5) shows us that while the stretch history $\mathbf{U}^t$ of a simple material may influence its present stress in any way whatever, past rotations have no influence at all. The present rotation $\mathbf{R}$ enters (4.5) explicitly. Thus, the reduced form enables us to dispense with considering rotation in determining the response of a motion. If we like, we may regard (4.3) as effecting an extension of $\mathfrak{G}$ from the range of positive-definite symmetric tensor histories to the full range of non-singular tensor histories.

The reduced form enables us also, in principle, to reduce the number of tests needed to determine the response functional $\mathfrak{G}$ by observation. Indeed, consider pure stretch histories: $\mathbf{R}^t = \mathbf{1}$. If we know the stress $\mathbf{T}$ corresponding to an arbitrary homogeneous pure stretch history $\mathbf{U}^t$, we have a relation of the form $\mathbf{T} = \mathfrak{G}(\mathbf{U}^t)$. By (4.5) we then know $\mathbf{T}$ for all deformation histories. Alternatively, consider irrotational histories: $\mathbf{W} = \mathbf{0}$. Given any $\mathbf{U}^t$, we can determine $\mathbf{R}^t$ by integrating $(2.19)_1$ with $\mathbf{W}$ set equal to $\mathbf{0}$. If we know the stress corresponding to an arbitrary irrotational history, by putting the corresponding values of $\mathbf{R}$ into (4.5) we can again determine $\mathfrak{G}$. Thus, we may characterize simple materials in either of two more economical ways: A material is simple if and only if its response to all deformations is determined by its response to all homogeneous pure stretch histories, or to all homogeneous irrotational deformation histories.

In the polar decomposition (2.1), two measures of stretch, $\mathbf{U}$ and $\mathbf{V}$, are introduced. Kinematically, there is no reason to prefer one to the other. From (4.3) we see that use of $\mathbf{U}$ to measure the stretch history leads to a simple reduced form for the constitutive equation of a simple material. If we like, of course we may use $\mathbf{V}$ instead. Since $\mathbf{U}^t = (\mathbf{R}^t)^T \mathbf{V}^t \mathbf{R}^t$, substitution in (4.3) shows that by using $\mathbf{V}$ we do not generally eliminate the rotation history $\mathbf{R}^t$. That is, use of $\mathbf{V}$ does not lead to a simple result. There are many other tensors that measure stretch just as well as $\mathbf{U}$ and $\mathbf{V}$. In the older literature one or another of these is called a "strain" tensor. I prefer not to use the term "strain" at all since the literature has been dirtied by futile, illusory, or even erroneous wrangling about the virtues of one or another strain measure. Here we shall simply use the results the equations bring us, not confusing ourselves with unnecessary words.

There are infinitely many other reduced forms for the constitutive equation of a simple material. Since $\underset{\sim}{C}^t = (\underset{\sim}{U}^t)^2$, one such form is

$$\begin{aligned} \underset{\sim}{T} &= \underset{\sim}{R}\underset{\sim}{U}\underset{\sim}{U}^{-1}\underset{\sim}{\mathfrak{G}}(\sqrt{\underset{\sim}{C}^t})\underset{\sim}{U}^{-1}\underset{\sim}{R}\underset{\sim}{R}^T \\ &= \underset{\sim}{F}\underset{\sim}{\mathfrak{L}}(\underset{\sim}{C}^t)\underset{\sim}{F}^T , \end{aligned} \tag{4.6}$$

where

$$\underset{\sim}{\mathfrak{L}}(\underset{\sim}{C}^t) \equiv \underset{\sim}{U}^{-1}\underset{\sim}{\mathfrak{G}}(\sqrt{\underset{\sim}{C}^t})\underset{\sim}{U}^{-1} .$$

Exercise 4.1. Using (1.40), derive NOLL's reduced form for the constitutive equation of a simple material:

$$\bar{\underset{\sim}{T}}(t) = \underset{\sim}{\mathfrak{K}}[\bar{\underset{\sim}{C}}_t^t ; \underset{\sim}{C}(t)] , \tag{4.7}$$

where for any tensor $\underset{\sim}{K}$ the tensor $\bar{\underset{\sim}{K}}$ is defined by

$$\bar{\underset{\sim}{K}} = \underset{\sim}{R}(t)^T \underset{\sim}{K}\underset{\sim}{R}(t) . \tag{4.8}$$

This last result, (4.7), shows that it is not possible to express the effect of the deformation history on the stress entirely by measuring deformation with respect to the present configuration. While the effect of all the past history, $0 < s < \infty$, is accounted for in this way, a fixed reference configuration is required, in general, to allow for the effect of the deformation at the present instant, as shown by the appearance of $C(t)$ as a parameter in (4.7). The result itself is important in that it enables us to go as far as possible toward avoiding use of a fixed reference configuration. Roughly, it shows that memory effects can be accounted for entirely by use of the relative deformation, but finite-strain effects require use of some fixed reference configuration, any one we please. This result should not surprise anyone, since in the theory of finite elastic strain the stress tensor is altogether independent of the relative deformation and hence cannot be expressed in terms of it.

Internal Constraints.

So far, we have been assuming that the material is capable, if subjected to appropriate forces, of undergoing any smooth motion. If the class of possible motions is limited at interior points of $\mathcal{B}$, the material is said to be subject to an internal constraint. A simple constraint is expressed by requiring the deformation gradient $\tilde{F}$ to satisfy an equation of the form

$$\gamma(\tilde{F}) = 0 , \tag{4.9}$$

where γ is a frame-indifferent scalar.

Exercise 4.2. Prove that $\gamma(\tilde{F})$ is frame-indifferent if and only if

$$\gamma(\tilde{F}) = \gamma(\tilde{U}) . \tag{4.10}$$

Hence a simple constraint may be written in the form

$$\lambda(\tilde{C}) = 0 , \tag{4.11}$$

where λ is a scalar function, and every such equation expresses a frame-indifferent simple constraint.

If we differentiate (4.11) with respect to time for a fixed particle, we obtain

$$\dot{\lambda} = \mathrm{tr}[\partial_{\tilde{C}}\lambda(\tilde{C})\dot{\tilde{C}}] = 0 . \tag{4.12}$$

But by (2.4) and (1.40),

$$\tilde{C}(\tau) = \tilde{F}(\tau)^T\tilde{F}(\tau) = \tilde{F}(t)^T\tilde{C}_t(\tau)\tilde{F}(t) . \tag{4.13}$$

If we differentiate this relation with respect to τ and then put $\tau = t$, by $(2.10)_2$ we find that

$$\dot{\tilde{C}}(t) = 2\tilde{F}(t)^T\tilde{D}(t)\tilde{F}(t) . \tag{4.14}$$

Hence (4.12) becomes

$$\mathrm{tr}[\tilde{F}\partial_{\tilde{C}}\lambda(\tilde{C})\tilde{F}^T\tilde{D}] = 0 \tag{4.15}$$

for all non-singular $\underset{\sim}{F}$ and all symmetric $\underset{\sim}{D}$. Conversely, if (4.15) holds at each instant for the particle in question, (4.12) follows from it by integration. Thus (4.15) may be used alternatively as a general expression for a frame-indifferent simple constraint.

Principle of Determinism for Constrained Simple Materials.

Constraints, since they consist in the prevention of some kinds of motion, must be maintained by forces. Since the constraints, by definition, are immutable, the forces maintaining them cannot be determined by the motion itself or its history. In particular, simple internal constraints must be maintained by appropriate stresses, and the constitutive equation of a constrained simple material must be such as to allow these stresses to operate, irrespective of the deformation history.

For constrained materials, accordingly, the principle of determinism must be relaxed. *A fortiori*, the necessary modification of that principle cannot be deduced from the principle itself but must be brought in as a new axiom.

There are, presumably, many systems of forces which could effect any given constraint. The simplest of these are the ones that do no work in any motion compatible with the constraint. In a constrained material these will therefore be assumed to remain arbitrary in the sense that they are not determined by the history of the motion.

Thus we have given reasons for laying down the following *Principle of determinism for simple materials subject to constraints*: The stress is determined by the history of the motion only to within an arbitrary tensor that does no work in any motion compatible with the constraint. That is,

$$\underset{\sim}{T} = \underset{\sim}{N} + \underset{\sim}{\mathfrak{G}}(\underset{\sim}{F}^t) , \tag{4.16}$$

where $\underset{\sim}{N}$ does no work in any motion satisfying the constraint, and where $\underset{\sim}{\mathfrak{G}}$ need be defined only for such arguments $\underset{\sim}{F}^t$ as satisfy the constraint. $\underset{\sim}{T} - \underset{\sim}{N}$ is called the *determinate stress*.

The problem is now to find $\underset{\sim}{N}$. The rate of working of a symmetric stress tensor $\underset{\sim}{T}$ in a motion with stretching tensor $\underset{\sim}{D}$ is the *stress working* w:

$$w \equiv tr(\underset{\sim}{T}\underset{\sim}{D}) . \tag{4.17}$$

Accordingly, we are to find the general solution $\tilde{N}$ of the equation

$$\operatorname{tr}(\tilde{N}\tilde{D}) = 0 \quad , \tag{4.18}$$

where $\tilde{D}$ is any symmetric tensor that satisfies (4.15). Now $\tilde{F}\partial_{\tilde{C}}\lambda(\tilde{C})\tilde{F}^T$ is a symmetric tensor, and the operation $\operatorname{tr}(\tilde{A}\tilde{B})$ defines an inner product in the space of symmetric tensors. Hence $\tilde{N}$, regarded as a vector, must be perpendicular to every vector $\tilde{D}$ that is perpendicular to $\tilde{F}\partial_{\tilde{C}}\lambda(\tilde{C})\tilde{F}^T$. Thus $\tilde{N}$ is parallel to this latter vector:

$$\tilde{N} = q\tilde{F}\partial_{\tilde{C}}\lambda(\tilde{C})\tilde{F}^T \quad , \tag{4.19}$$

where q is a scalar factor.

If there are k constraints $\lambda^i(\tilde{C}) = 0$, then

$$\tilde{N} = \sum_{i=1}^{k} q_i \tilde{F}\partial_{\tilde{C}}\lambda^i(\tilde{C})\tilde{F}^T \quad . \tag{4.20}$$

Substitution into (4.16) yields the general constitutive equation for simple material subject to k simple frame-indifferent internal constraints.

The determinate part of the stress, $\tilde{\mathfrak{G}}(\tilde{F}^t)$, may be expressed in reduced forms identical with those found for unconstrained materials.

Examples of Internal Constraints.

1. Incompressibility. A material is said to be incompressible if it can experience only isochoric motions. By $(1.22)_3$ and $(2.6)_9$, an appropriate constraint function is

$$\lambda(\tilde{C}) = \det \tilde{C} - 1 \quad . \tag{4.21}$$

Since

$$\tilde{F}\partial_{\tilde{C}}\lambda(\tilde{C})\tilde{F}^T = \tilde{F}\tilde{C}^{-1}\tilde{F}^T \det\tilde{C} = \tilde{1} \quad , \tag{4.22}$$

(4.19) yields

$$\tilde{N} = -p\tilde{1} \quad , \tag{4.23}$$

where p is an arbitrary scalar. Thus we have verified a result due in effect to POINCARÉ: In an incompressible material, the stress is determinate from the motion only to within an arbitrary hydrostatic pressure.

2. Inextensibility. If $\tilde{e}$ is a unit vector in the reference con-

figuration, $\mathbf{Fe}$ is the vector into which it is carried in a homogeneous deformation with gradient $\mathbf{F}$. Accordingly, for a material inextensible in the direction $\mathbf{e}$ an appropriate constraint function is

$$\lambda(\mathbf{C}) = (\mathbf{Fe})^2 - 1 = \mathbf{e}\cdot\mathbf{Ce} - 1 . \tag{4.24}$$

Since

$$\partial_{\mathbf{C}}\lambda(\mathbf{C}) = \mathbf{e} \otimes \mathbf{e} , \tag{4.25}$$

(4.19) yields

$$\mathbf{N} = q\mathbf{F}(\mathbf{e} \otimes \mathbf{e})\mathbf{F}^T = q\mathbf{Fe} \otimes \mathbf{Fe} . \tag{4.26}$$

Since $\mathbf{N}$ is an arbitrary uniaxial tension in the direction of $\mathbf{Fe}$, we recover a result first found by ADKINS & RIVLIN: In a material inextensible in a certain direction, the stress is determinate only to within a uniaxial tension in that direction.

3. Rigidity. A material is rigid if it is inextensible in every direction. By the result just established, the stress in a rigid material is determinate only to within an arbitrary tension in every direction. In other words, the stress in a rigid material is altogether unaffected by the motion.

Importance of Constrained Materials.

We have just made it plain that a constrained material is by no means a special case of an unconstrained one. Rather, the reverse holds, and the unconstrained material emerges as special. The behavior of a constrained material is not the same as that of an unconstrained one which happens to experience a motion satisfying the constraint. For example, if an unconstrained material happens to be subjected to an isochoric deformation history, its stress is determined by that history. An incompressible material, by definition, can never be subjected to anything but isochoric deformation histories, but its stress is never completely determined by them, being always indeterminate to the extent of an arbitrary hydrostatic pressure. Hydrodynamic writers are guilty of propagating not only bad English but also confusion when they refer to "incompressible flows". A flow, obviously, cannot be compressed. It may or may not be isochoric, and a fluid may or may not be incompressible; the behavior of a compressible fluid in an isochoric flow is generally not at all the same as that of an incompressible fluid in the same flow.

A constrained material is susceptible of a smaller class of deformations. Corresponding to this restriction, arbitrary stresses arise. Their presence makes a greater variety of response possible in those deformations that do occur. Consequently, solution of problems becomes easier.

The extreme case is furnished by the rigid material, where the deformations allowed are reduced to so special a class that the stress plays no part at all in the motion, which can be determined by solving ordinary differential equations which express no more than the principles of momentum and moment of momentum for the body as a whole.

The most useful case is that of the incompressible material. For a simple material in general, substitution of (3.25) into CAUCHY'S first law (3.16) yields when $\underset{\sim}{b} = \underset{\sim}{0}$

$$\operatorname{div} \underset{\sim}{\mathfrak{G}}(\underset{\sim}{F}^{t}) = \rho \ddot{\underset{\sim}{x}} \tag{4.27}$$

while for an incompressible material use of (4.23) and (4.16) in the same way yields

$$-\operatorname{grad} p + \operatorname{div} \underset{\sim}{\mathfrak{G}}(\underset{\sim}{F}^{t}) = \rho \ddot{\underset{\sim}{x}} \ . \tag{4.28}$$

These equations must be satisfied by any deformation history that can be produced by surface tractions alone. For the former, all $\underset{\sim}{F}^{t}$ are eligible to compete, and few will be found successful. For the latter, only those $\underset{\sim}{F}^{t}$ such that $\det \underset{\sim}{F}^{t} = 1$ are allowed, but p may be adjusted to aid in finding a solution. The condition upon the motion alone, of course, is now

$$\operatorname{curl} \operatorname{div} \underset{\sim}{\mathfrak{G}}(\underset{\sim}{F}^{t}) = \operatorname{curl} (\rho \ddot{\underset{\sim}{x}}) \ , \tag{4.29}$$

a differential equation of higher order than (4.27).

Nearly all the exact solutions found in non-linear continuum theories are for incompressible materials. In the next lecture we shall illustrate the distinction just made.

LECTURE 5: HOMOGENEOUS MOTIONS OF SIMPLE BODIES.

Significance of Homogeneous Motions.

The constitutive equation of a simple material,

$$\mathbf{T} = \mathfrak{G}_{\kappa}(\mathbf{F}^{(t)}(\mathbf{X},s);\ \mathbf{X}) , \qquad (3.25)$$

makes it plain that the response of a simple body to the class of all homogeneous motions, motions of the form

$$\mathbf{x} = \mathbf{F}(t)(\mathbf{X} - \mathbf{X}_0) + \mathbf{x}_0(t) , \qquad (3.24)$$

determines its response to all deformation histories. In an ideal program of experiment, then, we should subject a body of a given material to every deformation of the form (3.24) and record the stresses obtained. The results would amount to a determination of the response functional $\mathfrak{G}_{\kappa}$. We now ask whether such a program be possible in principle.

Can the motion (3.24) be produced in a body of the material defined by (3.25)? If the body force $\mathbf{b}$ in Cauchy's first law (3.16) is disposable, the answer is of course yes. However, while in considering the totality of dynamical processes we saw no reason to exclude any $\mathbf{b}$, it is a different matter when we come even to think about particular experiments, for only very special body forces are available in the laboratory. Practically speaking, a uniform field $\mathbf{b}$ = constant is all we are likely to be able to produce. We then ask if the homogeneous motion (3.24) can be produced in the material (3.25) if suitable surface tractions be supplied. We shall approach the problem only for homogeneous bodies.

Homogeneous Simple Bodies.

A simple body is homogeneous if there exists a reference configuration κ, called a homogeneous configuration, such that $\mathfrak{G}_{\kappa}$ does not depend explicitly on $\mathbf{X}$:

$$\mathbf{T} = \mathfrak{G}_{\kappa}[\mathbf{F}^{(t)}] . \qquad (5.1)$$

For a homogeneous simple body subject to a homogeneous strain history from a homogeneous configuration, then, $\mathbf{T} = \mathbf{T}(t)$, so that

$$\operatorname{div} \mathbf{T} = \mathbf{0} . \tag{5.2}$$

We have presumed so far that the body is not subject to any internal constraint. If, on the other hand, it is incompressible, a parallel definition and a parallel argument yield

$$\operatorname{div} \mathbf{T} = -\operatorname{grad} p , \tag{5.3}$$

where p is an arbitrary scalar field.

Solution for Unconstrained Materials.

The question we now put is, if $\mathbf{b} = \text{const.}$, is it possible to supply boundary tractions such as to produce the homogeneous motion (3.24) with respect to a homogeneous configuration of a simple body? Substitution of (5.2) in CAUCHY'S first law (3.16) yields the condition

$$\rho \mathbf{b} = \rho \ddot{\mathbf{x}} . \tag{5.4}$$

This requirement is compatible with the motion (3.24) if and only if

$$\ddot{\mathbf{F}} = \mathbf{0} , \quad \ddot{\mathbf{x}}_0 = \mathbf{b} . \tag{5.5}$$

Hence

$$\mathbf{F}(t) = \mathbf{F}_0(\mathbf{1} + t\mathbf{F}_1), \quad \mathbf{x}_0(t) = \frac{1}{2} t^2 \mathbf{b} + t\mathbf{e} + \mathbf{f} , \tag{5.6}$$

where $\mathbf{F}_0$ is an arbitrary non-singular tensor, $\mathbf{F}_1$ is an arbitrary tensor, $\mathbf{e}$ is an arbitrary vector, and $\mathbf{f}$ is an arbitrary fixed point.

Exercise 5.1. Prove that in these motions the velocity gradient is given by

$$\mathbf{G} = \mathbf{F}_0 \mathbf{F} \, (\mathbf{1} + t\mathbf{F}_1)^{-1} \mathbf{F}_0^{-1} . \tag{5.7}$$

From (5.6) we see that in general the homogeneous motion (3.24) cannot be produced by surface tractions alone when $\mathbf{b} = \text{const.}$ Only a special class of homogeneous motions, defined by (5.6), is possible. Therefore, the ideal program of determining $\mathfrak{G}_{\kappa}$ by effecting all homogeneous motions from κ

is not practicable. This does not mean that no other method of determining $\mathfrak{G}_\kappa$ may be found but merely that the method of homogeneous motions, used to motivate the definition of a simple material, is not feasible for finding $\mathfrak{G}_\kappa$ without use of artificial body forces.

What is left from our analysis is a particular class of homogeneous motions which can be effected in any unconstrained homogeneous simple body by application of suitable boundary tractions. In other words, a class of exact solutions for all homogeneous unconstrained bodies has been found. These particular solutions generally remain non-singular only for a finite interval of time. By assumption, $\det F(0) = \det F_0 \neq 0$. Then the motion is non-singular only so long as

$$\det (1 + tF_1) \neq 0 , \tag{5.8}$$

that is, in an interval of time (t_-, t_+) containing 0 and such that $-1/t$ never equals a proper number of F_1. Since F_1 is an arbitrary tensor, possibly singular, nothing can be said in general about its proper numbers. The possibilities that $t_- = -\infty$ or $t_+ = +\infty$ are not excluded. E.g., in an isochoric motion of this class,

$$|\det F_0| = 1 , \quad \det (1 + tF_1) = 1 , \tag{5.9}$$

and the interval in which the motion is non-singular is $(-\infty, +\infty)$.

Exercise 5.2. Prove that the motion is isochoric if and only if

$$|\det F_0| = 1 , \quad \mathrm{tr}F_1 = 0 , \quad \mathrm{tr}F_1^2 - (\mathrm{tr}F_1)^2 = 0 , \quad \det F_1 = 0 . \tag{5.10}$$

An important example is furnished by the motion of simple shearing, which has been used traditionally to illustrate every special theory of continuum mechanics. In a suitable pair of rectangular Cartesian systems, this motion has the form

$$x_1 = X_1$$

$$x_2 = X_2 + KtX_2 , \tag{5.11}$$

$$x_3 = X_3 .$$

Then

$$[\mathbf{F}] = \begin{Vmatrix} 1 & 0 & 0 \\ Kt & 1 & 0 \\ 0 & 0 & 1 \end{Vmatrix}, \qquad \mathbf{b} = \mathbf{0}, \tag{5.12}$$

so that

$$\mathbf{F}_0 = \mathbf{1}, \quad [\mathbf{F}_1] = K \begin{Vmatrix} 0 & 0 & 0 \\ 1 & 0 & 0 \\ 0 & 0 & 0 \end{Vmatrix}, \tag{5.13}$$

and (5.10) is satisfied. (In fact, $\mathbf{F}_1^2 = \mathbf{0}$.) Therefore, a simple shearing is possible, subject to the action of surface tractions alone, in any homogeneous simple body.

Another example is furnished by a homogeneous irrotational pure stretch: $\mathbf{R} = \mathbf{1}$, $\mathbf{W} = \mathbf{0}$, $\mathbf{U} = \mathbf{U}(t)$. By (2.19) $\mathbf{U}$ must satisfy the differential equation

$$\dot{\mathbf{U}}\mathbf{U} = \mathbf{U}\dot{\mathbf{U}}. \tag{5.14}$$

<u>Exercise 5.3</u>. Prove that (5.14) holds if and only if the proper vectors of $\mathbf{U}$ are constant in time. Thus

$$\mathbf{U} = \sum_{i=1}^{3} v_i(t)\mathbf{e}_i \otimes \mathbf{e}_i, \tag{5.15}$$

where the $\mathbf{e}_i$ are fixed mutually orthogonal unit vectors. Prove that the corresponding homogeneous motion has constant acceleration if and only if $\ddot{\mathbf{x}}_0 = \text{const.}$ and the v_i are positive linear functions of t. Prove that a rectangular block with faces normal to the $\mathbf{e}_i$ is transformed by the motion into another such block at any time within the interval for which the motion remains non-singular. Prove that this motion is isochoric if and only if it reduces to a state of rest.

The two families of motions just exhibited are particularly interesting special cases of the class of homogeneous motions we have proved to be possible in any homogeneous simple body by bringing to bear suitable boundary tractions alone. As we shall see, the corresponding class of solutions for incompressible bodies is much greater.

Preliminaries for Incompressible Bodies.

To study the corresponding problems for incompressible bodies, we shall allow a more general body force. If there exists a scalar $\upsilon(\underset{\sim}{x}, t)$ such that

$$\underset{\sim}{b} = -\text{grad } \upsilon \ , \tag{5.16}$$

the body force $\underset{\sim}{b}$ is said to be conservative. Cauchy's first law (3.16), for a homogeneous incompressible body subject to conservative body force, becomes

$$\text{div}(\underset{\sim}{T} + p\underset{\sim}{1}) - \rho\text{grad}\phi = \rho\ddot{\underset{\sim}{x}} \ , \tag{5.17}$$

where

$$\phi = \frac{p}{\rho} + \upsilon \ . \tag{5.18}$$

Now by (4.16) and (4.23), the quantity $\underset{\sim}{T} + p\underset{\sim}{1}$ is determined by the deformation history $\underset{\sim}{F}^t$. Suppose, then for a given history and given υ_1 a solution ϕ of (5.17) be known, and let the corresponding pressure be p_1. Consider another potential υ_2. If p_2 is so adjusted that

$$\frac{p_1}{\rho} + \upsilon_1 = \frac{p_2}{\rho} + \upsilon_2 \ , \tag{5.19}$$

then ϕ is unchanged, and so also is $\underset{\sim}{T} + p\underset{\sim}{1}$, since it is determined by $\underset{\sim}{F}^t$, and again we have a solution of (5.17). That is, if a given deformation history is possible in an incompressible material subject to one conservative body force, it is possible in that same material subject to any other conservative body force. In particular, we may take $\upsilon_2 = 0$ and conclude the following major result: In order for a motion to be possible in a given incompressible material, subject to the action of an arbitrary conservative body force and suitable boundary tractions, it is necessary and sufficient that it be possible subject to the action of boundary tractions alone. A statement to this effect is well known in Eulerian hydrodynamics; here it is seen to hold for all homogeneous incompressible bodies.

Let us now consider for a moment such motions as may be possible, subject to boundary tractions alone, in all homogeneous incompressible simple bodies. A motion of this kind must be possible, in particular, in an Eulerian incompressible fluid, which corresponds to the special case when $\mathfrak{G}(\mathbf{F}^t) \equiv \mathbf{0}$. Then

$$-\text{grad } p = \ddot{\mathbf{x}} \tag{5.20}$$

so the motion must be such that

$$\ddot{\mathbf{x}} = -\text{grad } \zeta \,, \tag{5.21}$$

where $\zeta = p + h(t)$, h being arbitrary. This result generalizes KELVIN'S theorem: In order to be possible in <u>all</u> homogeneous incompressible bodies, subject to boundary tractions alone, a motion must be circulation-preserving.[1] To any such motion, then, the classical hydrodynamic theorems of BERNOULLI, HELMHOLTZ, KELVIN, <u>etc</u>., apply.

Of course, the converse does not hold: Not all circulation-preserving motions can be produced in a typical homogeneous simple body by application of suitable surface tractions. Indeed, suppose that (5.21) holds. Then Cauchy's first law (3.16) yields

$$\text{div } \mathfrak{G}(\mathbf{F}^t) = -\rho \text{grad } \lambda \,, \tag{5.22}$$

where λ is a single-valued function. That is, the given motion is possible, subject to boundary tractions alone, if and only if

$$\text{curl div } \mathfrak{G}(\mathbf{F}^t) = \mathbf{0} \,, \tag{5.23}$$

---

[1] A motion $\mathbf{x} = \boldsymbol{\chi}(\mathbf{X},t)$ is <u>circulation-preserving</u> if for every circuit $\mathcal{C}$ in $\mathcal{B}$ the circulation around the configuration $\mathcal{C}_{\boldsymbol{\chi}}$ of $\mathcal{C}$ in $\mathcal{B}_{\boldsymbol{\chi}}$ is constant in time. That is,

$$\oint_{\mathcal{C}_{\boldsymbol{\chi}}} \mathbf{x} \cdot d\mathbf{x} = f(\mathcal{C}) \,.$$

The properties of circulation-preserving motions are explained in any good book on hydrodynamics and are developed from a general standpoint in pp. 105-138, <u>The Classical Field Theories</u>, Encyclopedia of Physics, Volume 3, Part 1, 1960.

and if the potential λ whose existence is assured by (5.23) be single-valued. If it is, then

$$\underset{\sim}{T} = -\rho(\zeta - \lambda - \upsilon + h)\,\underset{\sim}{1} + \underset{\sim}{\mathfrak{G}}(\underset{\sim}{F}^t) \;, \tag{5.24}$$

where $h = h(t)$ is an arbitrary function. This result follows at once by inserting (5.16), (5.21), and (5.22) into (3.16).

We have proved that a circulation-preserving motion (5.21) can be produced in the homogeneous incompressible body with response functional $\underset{\sim}{\mathfrak{G}}$ by boundary tractions alone if and only if (5.22) holds; the corresponding stress tensor is given by (5.24).

<u>Solution for Incompressible Materials</u>.

We may now apply the preceeding results to homogeneous motions. For them, (5.22) is always satisfied, and $\lambda = 0$. A condition that a motion be circulation-preserving is given by EULER'S criterion:

$$(\operatorname{grad} \ddot{\underset{\sim}{x}})^T = \operatorname{grad} \ddot{\underset{\sim}{x}} \;, \tag{5.25}$$

as may be seen at once from (5.21). In the case of the homogeneous motion (3.24),

$$\operatorname{grad} \ddot{\underset{\sim}{x}} = (\nabla\ddot{\underset{\sim}{x}})\underset{\sim}{F}^{-1} = \ddot{\underset{\sim}{F}}\underset{\sim}{F}^{-1} \;, \tag{5.26}$$

so that EULER'S criterion becomes

$$(\ddot{\underset{\sim}{F}}\underset{\sim}{F}^{-1})^T = \ddot{\underset{\sim}{F}}\underset{\sim}{F}^{-1} \;. \tag{5.27}$$

If this condition holds, it is easy to verify from (3.24) that (5.21) is indeed satisfied by the following potential:

$$-\zeta = (\underset{\sim}{x} - \underset{\sim}{x}_0) \cdot [\tfrac{1}{2}\ddot{\underset{\sim}{F}}\underset{\sim}{F}^{-1}(\underset{\sim}{x} - \underset{\sim}{x}_0) + \ddot{\underset{\sim}{x}}_0] \;. \tag{5.28}$$

We have proved, then, that a homogeneous isochoric motion is possible, subject to boundary tractions alone, in every incompressible simple body if and only if it satisfies (5.27). If (5.27) holds, we may obtain the stress by substitution in (5.24):

$$\mathbf{T} = \rho[(\mathbf{x} - \mathbf{x}_0) \cdot (\tfrac{1}{2}\ddot{\mathbf{F}}\mathbf{F}^{-1}(\mathbf{x} - \mathbf{x}_0) + \ddot{\mathbf{x}}_0) + \upsilon - h]\mathbf{1} + \mathfrak{G}(\mathbf{F}^t) \ . \qquad (5.29)$$

Any non-singular tensor $\mathbf{F}(t)$ that satisfies (5.27) and any point $\mathbf{x}_0(t)$ if put into (3.24) yield a possible motion, and substitution in (5.29) yields the stresses required to produce it, subject to the action of the body force (5.16).

In the previous lecture, we have shown that corresponding to any stretch $\mathbf{U}(t)$, we can determine a rotation $\mathbf{R}(t)$ such that $\mathbf{R}(t)\mathbf{U}(t)$ is irrotational. As is proved in works on classical hydrodynamics, every irrotational motion is circulation-preserving. If $\det \mathbf{U}(t) = 1$, the result demonstrated above shows that this motion can be produced in any homogeneous incompressible body by applying suitable boundary tractions. Consequently, the ideal experimental program proposed initially can be achieved, for homogeneous incompressible bodies, without calling upon artificial body forces, in fact without use of any body force at all.

This result can be obtained from a more general standpoint as follows.

Exercise 5.4. Prove that a motion is circulation-preserving if and only if it satisfies CAUCHY'S criterion:

$$\mathbf{F}^T\mathbf{W}\mathbf{F} = \mathbf{W}_0 = \text{const.} \qquad (5.30)$$

If we substitute CAUCHY'S criterion (5.30) into the identity $(2.19)_1$, we find that

$$\dot{\mathbf{R}} = \mathbf{R}\mathbf{Y} \qquad (5.31)$$

where

$$\mathbf{Y} = \frac{1}{2}(\mathbf{U}^{-1}\dot{\mathbf{U}} - \dot{\mathbf{U}}\mathbf{U}^{-1}) + \mathbf{U}^{-1}\mathbf{W}_0\mathbf{U}^{-1} \ . \qquad (5.32)$$

Suppose now a homogeneous stretch history $\mathbf{U}(t)$ and an arbitrary initial spin $\mathbf{W}_0$ be given. Then $\mathbf{Y}(t)$ is a known function. If $\mathbf{U}(t)$ and $\mathbf{W}_0$ be such that $\mathbf{Y}(t)$ is continuous, the first-order linear differential equation (5.31) determines a unique rotation $\mathbf{R}(t)$ corresponding to any assigned initial rotation $\mathbf{R}(0)$. The homogeneous motion with $\mathbf{F}(t) = \mathbf{R}(t)\mathbf{U}(t)$ is then circulation-preserving. By the above theorem on homogeneous motions, it follows that any homogeneous isochoric stretch history $\underset{\sim}{U}(t)$ determines for each choice of $\underset{\sim}{W}_0$ and $\underset{\sim}{R}(0)$ a unique homogeneous motion possible in <u>all</u> homogeneous incompressible simple bodies subject to boundary tractions alone. Conversely, these are the only homogeneous motions with the stretch $\underset{\sim}{U}(t)$ that are possible in <u>any one</u> homogeneous incompressible body without bringing to bear a non-conservative body force. The earlier assertion regarding irrotational motions corresponds to the special case when $\underset{\sim}{W}_0 = \underset{\sim}{0}$.

<u>Exercise 5.5</u>. Prove that a pure stretch $\underset{\sim}{F}(t) = \underset{\sim}{U}(t)$ is circulation-preserving if and only if

$$\dot{\underset{\sim}{U}}\underset{\sim}{U} - \underset{\sim}{U}\dot{\underset{\sim}{U}} = \text{const.} , \qquad (5.33)$$

generalizing (5.14). Hence, in general, a homogeneous isochoric pure stretch cannot be produced in any homogeneous incompressible simple body by the effect of boundary tractions alone. Among those special homogeneous isochoric pure stretches that can be so produced are the irrotational ones, given by (5.15) with the added restriction $v_1(t)v_2(t)v_3(t) = 1$.

<u>Comments</u>.

The results obtained in this lecture show the difference between the stress system in a compressible body that just happens to undergo an isochoric deformation and that in an incompressible body undergoing the same deformation. For the unconstrained body, change of volume is avoided because the stresses are selected in just the right way, and that way is specified uniquely by the constitutive functional. For the incompressible body, no system of stresses can produce anything but an isochoric motion, and corresponding to that fact there is an arbitrary hydrostatic pressure, unaffected by the motion.

Internal constraints such as incompressibility reduce the class of possible motions but expand the class of stresses compatible with such motions as may take place. The theory of a constrained material is therefore essentially easier to work out. The far-reaching simplification that results from assuming the material to be incompressible was seen and exploited by RIVLIN in his early researches on the non-linear continuum theories. Most of the explicit solutions known today concern incompressible bodies.

LECTURE 6: THE ISOTROPY GROUP. SOLIDS, ISOTROPIC MATERIALS, FLUIDS, FLUID CRYSTALS.

Isotropy Group

In the older literature a material is said to be "isotropic" if it is "unaffected by rotations." This means that if we first rotate a specimen of material and then do an experiment upon it, the outcome is the same as if the specimen had not been rotated. In other words, within the class of effects considered by the theory, rotations cannot be detected by any experiment. The response of the material with respect to the reference configuration $\underset{\sim}{\kappa}$ is the same as that with respect to any other obtained from it by rotation.

NOLL's concept of isotropy group (1958) generalizes this old idea. Rather than lay down a statement of material symmetry, we consider the material defined by the response functional $\underset{\sim}{\mathfrak{G}}_{\underset{\sim}{\kappa}}$ and ask what symmetries that material may have. Of course, those symmetries will depend on $\underset{\sim}{\kappa}$. In Lecture 3 we have derived an equation for the change of response functional under change of reference configuration:

$$\underset{\sim}{\mathfrak{G}}_{\underset{\sim}{\kappa}_2}(\underset{\sim}{F}^t) = \underset{\sim}{\mathfrak{G}}_{\underset{\sim}{\kappa}_1}(\underset{\sim}{F}^t\underset{\sim}{P}) \quad , \tag{3.28}$$

where $\underset{\sim}{P} = \nabla\underset{\sim}{\lambda}$, and $\underset{\sim}{\lambda}$ is the mapping from $\underset{\sim}{\kappa}_1$ to $\underset{\sim}{\kappa}_2$. This formula asserts that $\underset{\sim}{\mathfrak{G}}_{\underset{\sim}{\kappa}_2}$ and $\underset{\sim}{\mathfrak{G}}_{\underset{\sim}{\kappa}_1}$, if they are related in this way, describe the response of the same material to the deformation history $\underset{\sim}{F}^t$, in the one case when $\underset{\sim}{\kappa}_1$ is used as the reference configuration, in the other case when $\underset{\sim}{\kappa}_2$ is. We may consider (3.28) analogous to a transformation in analytic geometry, where the same figure is described by different equations in different co-ordinate systems.

We now ask, under what conditions does $\underset{\sim}{P}$ lead from a given $\underset{\sim}{\kappa}_1$ to a reference configuration $\underset{\sim}{\kappa}_2$ which is indistinguishable from it by experiments relating the stress to the deformation history $\underset{\sim}{F}^t$? First, simple experience in Eulerian hydrodynamics tells us that a change of density leads to different response, so we shall consider only those configurations with the same density.

By (1.18) and (1.19), then, we restrict attention to unimodular tensors $\underset{\sim}{H}$: tensors such that $\det \underset{\sim}{H} = \pm 1$. Suppose that $\underset{\sim}{H}$ is a unimodular tensor that satisfies the equation

$$\underset{\sim}{\mathcal{G}}_{\underset{\sim}{\kappa}_1}(\underset{\sim}{F}^t \underset{\sim}{H}) = \underset{\sim}{\mathcal{G}}_{\underset{\sim}{\kappa}_1}(\underset{\sim}{F}^t) \tag{6.1}$$

for every non-singular $\underset{\sim}{F}^t$. Note that $\underset{\sim}{\kappa}_1$ appears on both sides here. By (3.28), for all $\underset{\sim}{F}^t$,

$$\underset{\sim}{T} = \underset{\sim}{\mathcal{G}}_{\underset{\sim}{\kappa}_1}(\underset{\sim}{F}^t) = \underset{\sim}{\mathcal{G}}_{\underset{\sim}{\kappa}_2}(\underset{\sim}{F}^t) , \tag{6.2}$$

where $\underset{\sim}{\kappa}_2$ is the reference configuration obtained from $\underset{\sim}{\kappa}_1$ by the transformation $\underset{\sim}{\lambda}$ such that $\underset{\sim}{H} = \nabla\underset{\sim}{\lambda}$. Thus $\underset{\sim}{H}$ is the gradient of a deformation that maps $\underset{\sim}{\kappa}_1$ onto another configuration $\underset{\sim}{\kappa}_2$ from which any deformation history $\underset{\sim}{F}^t$ leads to just the same stress as it does from $\underset{\sim}{\kappa}_1$. In other words, $\underset{\sim}{H}$ corresponds to a density-preserving transformation which cannot be detected by any experiment.

Suppose both $\underset{\sim}{H}_1$ and $\underset{\sim}{H}_2$ satisfy (6.1). Then by two applications of (6.1)

$$\underset{\sim}{\mathcal{G}}_{\underset{\sim}{\kappa}_1}(\underset{\sim}{F}^t \underset{\sim}{H}_1 \underset{\sim}{H}_2) = \underset{\sim}{\mathcal{G}}_{\underset{\sim}{\kappa}_1}(\underset{\sim}{F}^t \underset{\sim}{H}_1) = \underset{\sim}{\mathcal{G}}_{\underset{\sim}{\kappa}_1}(\underset{\sim}{F}^t) . \tag{6.3}$$

Hence $\underset{\sim}{H}_1 \underset{\sim}{H}_2$ is a solution. If $\underset{\sim}{H}$ is a particular unimodular tensor, $\underset{\sim}{F}^t \underset{\sim}{H}^{-1}$ is a non-singular deformation history for every non-singular $\underset{\sim}{F}^t$, and any non-singular deformation history may be written in the form $\underset{\sim}{F}^t \underset{\sim}{H}^{-1}$. Replacing $\underset{\sim}{F}^t$ by $\underset{\sim}{F}^t \underset{\sim}{H}^{-1}$ in (6.1) shows that $\underset{\sim}{H}^{-1}$ is also a solution if $\underset{\sim}{H}$ is. Hence the totality of all solutions $\underset{\sim}{H}$ forms a group, the <u>isotropy group</u> $g_{\underset{\sim}{\kappa}}$ of the material at the particle whose place in the reference configuration $\underset{\sim}{\kappa}$ is $\underset{\sim}{X}$. From its definition, $g_{\underset{\sim}{\kappa}}$ is a subgroup of the unimodular group $\mathcal{U}$:

$$g_{\underset{\sim}{\kappa}} \subset \mathcal{U} . \tag{6.4}$$

The isotropy group is the collection of all static density-preserving deformations from $\underset{\sim}{\kappa}$ at $\underset{\sim}{X}$ that cannot be detected by experiment. Alternatively we may describe the isotropy group as the group of material symmetries. We have proved that every material has a non-empty isotropy group for every $\underset{\sim}{\kappa}$ and $\underset{\sim}{X}$.

Orthogonal Part of the Isotropy Group.

The members $\underset{\sim}{H}$ of $g_{\underset{\sim}{\kappa}}$ need not be orthogonal, but they may be. If an orthogonal tensor $\underset{\sim}{Q}\epsilon g_{\underset{\sim}{\kappa}}$, then also $\underset{\sim}{Q}^T\epsilon g_{\underset{\sim}{\kappa}}$, since $g_{\underset{\sim}{\kappa}}$ is a group and $\underset{\sim}{Q}^T = \underset{\sim}{Q}^{-1}$, so (6.1) yields

$$\underset{\sim}{\mathfrak{G}}_{\underset{\sim}{\kappa}}(\underset{\sim}{Q}\underset{\sim}{F}^t) = \underset{\sim}{\mathfrak{G}}_{\underset{\sim}{\kappa}}(\underset{\sim}{Q}\underset{\sim}{F}^t\underset{\sim}{Q}^T) \quad . \tag{6.5}$$

But in (4.1), which expresses the principle of material frame-indifference, we may choose the history $\underset{\sim}{Q}^t(s) = \underset{\sim}{Q}$ for all s and t and obtain

$$\underset{\sim}{\mathfrak{G}}_{\underset{\sim}{\kappa}}(\underset{\sim}{Q}\underset{\sim}{F}^t) = \underset{\sim}{Q}\,\underset{\sim}{\mathfrak{G}}_{\underset{\sim}{\kappa}}(\underset{\sim}{F}^t)\underset{\sim}{Q}^T . \tag{6.6}$$

Hence $\underset{\sim}{Q}\epsilon g_{\underset{\sim}{\kappa}}$ only if

$$\underset{\sim}{\mathfrak{G}}_{\underset{\sim}{\kappa}}(\underset{\sim}{Q}\underset{\sim}{F}^t\underset{\sim}{Q}^T) = \underset{\sim}{Q}\,\underset{\sim}{\mathfrak{G}}_{\underset{\sim}{\kappa}}(\underset{\sim}{F}^t)\underset{\sim}{Q}^T \tag{6.7}$$

for all non-singular deformation histories $\underset{\sim}{F}^t$. Conversely, suppose (6.7) holds for a particular orthogonal tensor $\underset{\sim}{Q}$ and for all non-singular $\underset{\sim}{F}^t$. In (6.6), which is a corollary to the principle of material frame-indifference, we may replace $\underset{\sim}{F}^t$ by $\underset{\sim}{F}^t\underset{\sim}{Q}^t$ and by use of (6.7) conclude that $\underset{\sim}{Q}^T$ is a member of $g_{\underset{\sim}{\kappa}}$. Thus (6.7) is the equation to be solved to find all orthogonal members of $g_{\underset{\sim}{\kappa}}$.

Clearly the central inversion, $-\underset{\sim}{1}$, satisfies (6.7). Hence $-\underset{\sim}{1}\epsilon g_{\underset{\sim}{\kappa}}$ always. The central inversion cannot be visualized as a deformation of the material but corresponds to a change from a right-handed to a left-handed frame of reference, or vice-versa. We have shown, then, that an inversion of the reference configuration cannot be detected. (In a more general theory, involving e.g. heat conduction or electromagnetism, an isotropy group may be defined, but $-\underset{\sim}{1}$ generally will fail to be a member of it.)

Since $\underset{\sim}{1}$ and $-\underset{\sim}{1}$ from a group by themselves, that group is the smallest possible isotropy group:

$$\{\underset{\sim}{1}, -\underset{\sim}{1}\} \subset g_{\underset{\sim}{\kappa}} \subset u \quad . \tag{6.8}$$

Change of Reference Configuration.

Exercise 6.1. Prove NOLL's rule:

$$g_{\underset{\sim}{\kappa}_2} = \underset{\sim}{P} g_{\underset{\sim}{\kappa}_1} \underset{\sim}{P}^{-1} . \tag{6.9}$$

The isotropy group $g_{\underset{\sim}{K}}$ depends upon the reference configuration, since the symmetries a material enjoys with respect to one configuration generally differ from those it enjoys with respect to another. NOLL'S rule shows that if $g_{\underset{\sim}{\kappa}}$ is known for one configuration, it is known for all and may be calculated explicitly. That is, the symmetries of a material in any one configuration determine its symmetries in every other. Notice that $\underset{\sim}{P}$ in (6.9) need not be unimodular.

From NOLL's rule we see that if $\underset{\sim}{P} = \alpha\underset{\sim}{1}$, where $\alpha \neq 0$, or if $\underset{\sim}{P} = -\underset{\sim}{1}$, $g_{\underset{\sim}{\kappa}}$ is unchanged. That is, inversions and dilatations leave material symmetries invariant.

Moreover, there are two particular groups which are invariant under (6.9). First, obviously,

$$\underset{\sim}{P}\{\underset{\sim}{1}, -\underset{\sim}{1}\}\underset{\sim}{P}^{-1} = \{\underset{\sim}{1}, -\underset{\sim}{1}\} . \tag{6.10}$$

Second, with $\underset{\sim}{P}$ given, $\underset{\sim}{P}\underset{\sim}{H}\underset{\sim}{P}^{-1}$ is unimodular for all unimodular $\underset{\sim}{H}$, and if $\bar{\underset{\sim}{H}}$ is any unimodular tensor, the tensor $\underset{\sim}{P}^{-1}\bar{\underset{\sim}{H}}\underset{\sim}{P}$ is a unimodular tensor $\underset{\sim}{H}$, and $\bar{\underset{\sim}{H}} = \underset{\sim}{P}\underset{\sim}{H}\underset{\sim}{P}^{-1}$. Therefore

$$\underset{\sim}{P}\,u\,\underset{\sim}{P}^{-1} = u . \tag{6.11}$$

That is, if $g_{\underset{\sim}{\kappa}} = \{\underset{\sim}{1}, -\underset{\sim}{1}\}$ for one $\underset{\sim}{\kappa}$, it does so for all, and likewise if $g_{\underset{\sim}{\kappa}} = u$. The smallest isotropy group, viz $\{\underset{\sim}{1}, -\underset{\sim}{1}\}$, corresponds to a material which has no non-trivial symmetries. The result (6.10) shows that no deformation can bring this material into a configuration which does have non-trivial symmetries. In this material, then, deformation cannot produce any symmetry. Likewise, by (6.11) in a material with maximum symmetry, $g = u$, and no deformation can destroy this symmetry.

Isotropic Materials.

The concept of isotropy is one of CAUCHY's greatest discoveries. In the present framework, it corresponds precisely to the following definition: A material is isotropic if there exists a reference configuration κ such that

$$g_{\kappa} \supset \mathcal{O} \quad , \tag{6.12}$$

where $\mathcal{O}$ is the orthogonal group. Such a reference configuration is called an undistorted state of the isotropic material. According to the definition, no orthogonal deformation from an undistorted state is discernible. By NOLL's rule (6.11), any orthogonal transformation carries one undistorted state into another.

For an isotropic material (6.7) changes from an equation to be solved for certain Q into an identity satisfied by all Q, and likewise (6.1) is satisfied by all orthogonal H. By (6.1), then, the value of T is unchanged if we replace F^t by F^tQ, where Q is any constant orthogonal tensor. At any given time t, we may thus replace $F^t(s)$ by $F^t(s)Q(t)$, and in particular by $F^t(s)R(t)^T$, without changing the value of T. We apply this transformation to the constitutive equation in the form (4.7). In this transformation $F(t)$ is replaced by $F(t)R(t)^T$, and hence $R(t)$ is replaced by 1; $C(t)$ is replaced by $R(t)C(t)R(t)^T$, which by virtue of (2.4) is $B(t)$; and, by (1.40) and (4.8), $\bar{C}_t^t$ is replaced by C_t^t. Thus (4.7) yields as NOLL'S reduced form of the constitutive equation for isotropic materials.

$$T = \mathfrak{H}[C_t^t \; ; \; B(t)] \; , \tag{6.13}$$

in which, as is to be expected, the rotation does not appear at all.

According to (6.7), moreover, if $F^t(s)$ is replaced by $Q(t)F^t(s)Q(t)^T$, for any $Q(t)$, the stress $T(t)$ is replaced by $Q(t)T(t)Q(t)^T$. In this replacement, $C_t^t(s)$ and $B(t)$ are replaced by $Q(t)C_t^t(s)Q(t)^T$ and $Q(t)B(t)Q(t)^T$, as is easily verified from (4.8) and (1.40). Thus the functional $\mathfrak{H}$ in (6.13) must satisfy the identity

$$\mathfrak{H}[QC_t^tQ^T \; ; \; QB(t)Q^T] = Q\,\mathfrak{H}[C_t^t \; ; \; B(t)]Q^T \; . \tag{6.14}$$

(Notice that here $\underset{\sim}{Q} = \underset{\sim}{Q}(t)$, a function of the present time, not a history.) A functional satisfying this requirement for all $\underset{\sim}{Q}$ is called isotropic. Conversely, if (6.14) is satisfied by $\underset{\sim}{\mathfrak{R}}$, (6.13) gives the constitutive equation of an isotropic simple material, referred to an undistorted configuration. If reference configurations that are not undistorted are used, the constitutive equation of an isotropic material cannot have the form (6.13) and generally shows no recognizable simplicity.

Exercise 6.2. Verify the steps in the foregoing proof.

While (6.12) embodies a natural concept of isotropy, it seems more general than in fact it is. According to a theorem of group theory, the orthogonal group is maximal in the unimodular group. That is, if $\mathcal{g}$ is a group such that $\mathcal{O} \subset \mathcal{g} \subset \mathcal{U}$, then

$$\text{either } \mathcal{g} = \mathcal{O} \text{ or } \mathcal{g} = \mathcal{U} . \qquad (6.15)$$

Thus the isotropy group of an isotropic material in an undistorted state is either the orthogonal group or the unimodular group.

Solids.

A material is thought of as "solid" if it has some "preferred configuration", a change of shape from which changes some of its properties. A change of shape is a non-orthogonal deformation. Thus a solid has some configuration from which any non-orthogonal deformation is detectable by subsequent experiments. Formally, then, a simple material is a solid if there exists a reference configuration $\underset{\sim}{\kappa}$ such that

$$\mathcal{g}_{\underset{\sim}{\kappa}} \subset \mathcal{O} . \qquad (6.16)$$

Such a $\underset{\sim}{\kappa}$ is called an undistorted state. According to this definition, no non-orthogonal transformation belongs to the isotropy group $\mathcal{g}_{\underset{\sim}{\kappa}}$ when $\underset{\sim}{\kappa}$ is an undistorted state.

The material for which $\mathcal{g} = \{\underset{\sim}{1}, -\underset{\sim}{1}\}$ is a solid. Indeed, it is called triclinic, and it furnishes an example of a crystalline solid in the classical sense. All the classical crystallographic groups, provided they be extended so as to include $-\underset{\sim}{1}$, correspond to solids. So also do the groups defining "transversely isotropic" and "orthotropic" materials, and many others.

For solids, no particularly simple form of the constitutive equation has been found.

For an isotropic solid, by hypothesis, there exist configurations $\underset{\sim}{\kappa}$ and $\bar{\underset{\sim}{\kappa}}$ such that

$$\mathcal{g}_{\underset{\sim}{\kappa}} \supset \mathcal{O} \qquad \mathcal{g}_{\bar{\underset{\sim}{\kappa}}} \subset \mathcal{O} \ . \tag{6.17}$$

(Here $\underset{\sim}{\kappa}$ is an undistorted state in the sense of "isotropic material", while $\bar{\underset{\sim}{\kappa}}$ is an undistorted state of the solid.) By (6.15), either $\mathcal{g}_{\underset{\sim}{\kappa}} = \mathcal{O}$ or $\mathcal{g}_{\underset{\sim}{\kappa}} = \mathcal{U}$. If $\mathcal{g}_{\underset{\sim}{\kappa}} = \mathcal{U}$, then by (6.11) $\mathcal{g}_{\bar{\underset{\sim}{\kappa}}} = \mathcal{U}$, contradicting $(6.17)_2$. Hence $\mathcal{g}_{\underset{\sim}{\kappa}} = \mathcal{O}$. In particular, $\underset{\sim}{\kappa}$ is also an undistorted state of the solid.

<u>Exercise 6.3</u>. Let $\underset{\sim}{\kappa}$ and $\underset{\sim}{\kappa}^*$ be two undistorted states of a solid, and let $\underset{\sim}{P} = \underset{\sim}{\nabla}\underset{\sim}{\lambda}$, where $\underset{\sim}{\lambda}: \underset{\sim}{\kappa} \to \underset{\sim}{\kappa}^*$. If the polar decomposition of $\underset{\sim}{P}$ is $\underset{\sim}{P} = \underset{\sim}{R}\underset{\sim}{U}$, prove that

$$\mathcal{g}_{\underset{\sim}{\kappa}^*} = \underset{\sim}{R}\mathcal{g}_{\underset{\sim}{\kappa}}\underset{\sim}{R}^{-1} \ . \tag{6.18}$$

That is, $\mathcal{g}_{\underset{\sim}{\kappa}^*}$ is an orthogonal conjugate of $\mathcal{g}_{\underset{\sim}{\kappa}}$.

By (6.17) and (6.18), $\mathcal{g}_{\bar{\underset{\sim}{\kappa}}}$ is an orthogonal conjugate of $\mathcal{O}$. The only such conjugate is $\mathcal{O}$. Hence $\mathcal{g}_{\underset{\sim}{\kappa}} = \mathcal{O}$. We have proved, then, that <u>the isotropy group of an isotropic solid in any undistorted state</u> $\underset{\sim}{\kappa}$ <u>is the orthogonal group</u>:

$$\mathcal{g}_{\underset{\sim}{\kappa}} = \mathcal{O} \ . \tag{6.19}$$

<u>Fluids</u>.

There are various physical motions concerned with fluids. One is that a fluid is a substance which can flow. "Flow" is itself a vague term. One meaning of "flow" is simply deformation under stress, which does not distinguish a fluid from a solid. Another is that steady velocity results from constant stress, which seems to be special and inapplicable except to particular cases. Another is the inability to support shear stress when in equilibrium. Formally, within the theory of simple materials, such a definition would yield

$$\underset{\sim}{T} = -p(\rho)\underset{\sim}{1} + \underset{\sim}{\mathfrak{F}}(\underset{\sim}{F}^t) \ , \tag{6.20}$$

where $\underset{\sim}{\mathfrak{C}}(\underset{\sim}{1}) = \underset{\sim}{0}$. Since the material so defined may have any isotropy group whatever, including one of those already considered to define a solid, this definition does not lend itself to a criterion in terms of material symmetry. Finally, a fluid is regarded as a material having no preferred configurations. In terms of isotropy groups, $g_{\underset{\sim}{\kappa}_1} = g_{\underset{\sim}{\kappa}_2}$ for all $\underset{\sim}{\kappa}_1$ and $\underset{\sim}{\kappa}_2$.

Accordingly, we adopt the following definition: <u>A fluid is a non-solid material with no preferred configurations</u>. By NOLL's rule (6.9), the isotropy group g of a fluid satisfies

$$g = \underset{\sim}{H} g \underset{\sim}{H}^{-1} \quad , \tag{6.21}$$

for all unimodular $\underset{\sim}{H}$. We have seen already that $g = \{\underset{\sim}{1}, -\underset{\sim}{1}\}$ and $g = u$ satisfy this equation. A theorem in group theory asserts that u is a "simple group", which means that there are no solutions of (6.21) beyond the two trivial ones just specified. The former solution corresponds to a solid. Hence the only group compatible with the definition is $g = u$: <u>The isotropy group of a fluid, in every configuration, is the unimodular group</u>. As a corollary, <u>every fluid is isotropic</u>, and every configuration of a fluid is undistorted.

Since fluids are isotropic, we can apply (6.13). Since for a fluid $\underset{\sim}{T}$ cannot be changed by a static deformation from one configuration to another with the same density, the dependence upon $\underset{\sim}{B}(t)$ must reduce to dependence on $\det \underset{\sim}{B}(t)$, or, what is the same thing, dependence on ρ:

$$\underset{\sim}{T} = \mathfrak{H}(\underset{\sim}{C}_t^t\ ; \rho) \quad . \tag{6.22}$$

Thus the stress in a fluid, at a given density, is determined by the history of the relative deformation alone. Of course, $\mathfrak{H}$ must satisfy (6.14). For the particular case of the rest history, $\underset{\sim}{C}_t^t = \underset{\sim}{1}$, so that (6.14) yields

$$\underset{\sim}{T} = \underset{\sim}{\mathfrak{H}}(\underset{\sim}{1};\ \rho) = \underset{\sim}{Q}\underset{\sim}{\mathfrak{H}}(\underset{\sim}{1}, \rho)\underset{\sim}{Q}^T = \underset{\sim}{Q}\underset{\sim}{T}\underset{\sim}{Q}^T \ . \tag{6.23}$$

That is, $\underset{\sim}{T}$ in a fluid at rest commutes with every orthogonal tensor. The only tensor satisfying this requirement is $\underset{\sim}{T} = -p(\rho)\underset{\sim}{1}$. Therefore the stress in a fluid at rest is a hydrostatic pressure which depends on the density alone, and (6.22) may be put into the alternative form

$$\underset{\sim}{T} = -p(\rho)\underset{\sim}{1} + \underset{\sim}{\mathfrak{F}}(\underset{\sim}{K}_t^t ; \rho), \tag{6.24}$$

where

$$\underset{\sim}{K}_t^t(s) \equiv \underset{\sim}{C}_t^t(s) - \underset{\sim}{1} , \tag{6.25}$$

$$\underset{\sim}{\mathfrak{F}}(\underset{\sim}{0}; \rho) = \underset{\sim}{0} .$$

This result constitutes <u>NOLL's fundamental theorem</u> on fluids. It asserts, among other things, that a fluid is a substance which can flow in the sense expressed by (6.20).

A fluid may react to its entire deformation history, yet such reaction cannot be different in different configurations with the same density. A fluid reconciles these two seemingly contradictory qualities - ability to remember all its past and inability to regard one configuration as different from another - by reacting to the past only in so far as it differs from the ever-changing present. This statement is the content of NOLL's theorem.

<u>Fluid Crystals</u>.

To exhaust the possible types of simple materials, we define any non-solid as being a <u>fluid crystal</u>. For a fluid crystal, then, there exists no reference configuration $\underset{\sim}{\kappa}$ such that $\mathcal{g}_{\underset{\sim}{\kappa}} \subset \mathcal{O}$. Of course, every isotropy group has orthogonal elements. For a fluid crystal, the isotropy group with respect to every configuration has some non-orthogonal elements. That is, from every configuration of a fluid crystal there is <u>some</u> undetectable change of shape. In this regard a fluid crystal is like a fluid, for which all changes of shape without change of density are undetectable. Since it is impossible that $\mathcal{g}_{\underset{\sim}{\kappa}} \supset \mathcal{O}$ unless the fluid crystal is in fact a fluid, for an isotropic fluid crystal there are also some rotations that are detectable. Clearly <u>a fluid crystal is a fluid if and only if it is isotropic</u>.

LECTURE 7. MOTIONS WITH CONSTANT STRETCH HISTORY.

Definition.

The simple material may have a long memory. Deformations undergone an arbitrarily long time ago may continue to affect and alter the present stress. With such complicated material response possible, the solution of particular problems may be extremely difficult, indeed unfeasible. There are particular motions, however, in which memory effects are given little chance to manifest themselves, because there is little to remember.

Consider, for example, the constitutive equation of a simple fluid in the form (6.22):

$$\tilde{T} = \tilde{\mathfrak{R}}(\tilde{C}_t^t;\ \rho). \tag{6.22}$$

In the particular case when ρ = const. and $\tilde{C}_t^t(s)$ is the same function of s for all t, the stress becomes constant in time for a given particle. The particle may have experienced deformation for all past time, but as it looks backward it sees the entire sequence of past deformations relative to its present configuration remain unchanged.

More generally, in view of the principle of frame-indifference, essentially the same simplification will result if there exists an orthogonal tensor $\tilde{Q}(t)$ such that

$$\tilde{C}_t^t(s) = \tilde{Q}(t)\tilde{C}_0^0(s)\tilde{Q}(t)^T. \tag{7.1}$$

Here, of course, $\tilde{C}_0^0$ denotes $\tilde{C}_t^t$ when $t = 0$. Such motions were introduced and called <u>substantially stagnant</u> by COLEMAN. In them, an observer situated upon the moving particle may orient himself in such a way as to see behind him always the same deformation history relative to the present configuration. The proper numbers of $\underset{\sim}{C}^{t}_{t}(s)$ are the same as those of $\underset{\sim}{C}^{0}_{0}(s)$, although the principal axes of the one tensor may rotate arbitrarily with respect to those of the other. Thus the principal relative stretches $v_{(t)i}$ remain the same functions of time, measured back from the present instant:

$$\overset{t}{v}_{(t)i}(s) = \overset{0}{v}_{(0)i}(s). \tag{7.2}$$

The functions $\overset{0}{v}_{(0)i}(s)$, of course, may be arbitrary. By (7.2), substantially stagnant motions may be defined alternatively as those having <u>constant stretch history</u>.

<u>NOLL's Theorem</u>.

All such motions are characterized by a <u>fundamental theorem</u> of <u>NOLL</u>: A motion has constant stretch history if and only if there exist an orthogonal tensor $\underset{\sim}{Q}(t)$, a scalar κ, and a constant tensor $\underset{\sim}{N}_0$ such that

$$\begin{aligned} \underset{\sim}{F}_0(\tau) &= \underset{\sim}{Q}(\tau)e^{\tau\kappa\underset{\sim}{N}_0}, \\ \underset{\sim}{Q}(0) &= \underset{\sim}{1}, \qquad |\underset{\sim}{N}_0| = 1. \end{aligned} \tag{7.3}$$

To prove NOLL's theorem, we begin from the hypothesis (7.1) and set

$$\underset{\sim}{H}(s) \equiv \underset{\sim}{C}_0(-s) = \underset{\sim}{Q}(t)^T\underset{\sim}{C}_t(t - s)\underset{\sim}{Q}(t). \tag{7.4}$$

By (1.41), $\underset{\sim}{F}_t(\tau) = \underset{\sim}{F}_0(\tau)\underset{\sim}{F}_0(t)^{-1}$, so

$$\begin{aligned} \underset{\sim}{Q}(t)\underset{\sim}{H}(s)\underset{\sim}{Q}(t)^T &= \underset{\sim}{C}_t(t - s), \\ &= [\underset{\sim}{F}_0(t)^T]^{-1}\underset{\sim}{C}_0(t - s)\underset{\sim}{F}_0(t)^{-1}, \\ &= [\underset{\sim}{F}_0(t)^T]^{-1}\underset{\sim}{H}(s - t)\underset{\sim}{F}_0(t)^{-1}. \end{aligned} \tag{7.5}$$

If we set

$$\underset{\sim}{E}(t) \equiv \underset{\sim}{Q}(t)^T\underset{\sim}{F}_0(t), \tag{7.6}$$

then (7.5) assumes the form of a difference equation:

$$\mathbf{H}(s - t) = \mathbf{E}(t)^T\mathbf{H}(s)\mathbf{E}(t). \tag{7.7}$$

To obtain a necessary condition for a solution $\mathbf{H}(s)$, we differentiate (7.7) with respect to t and put $t = 0$, obtaining the first-order linear differential equation

$$-\dot{\mathbf{H}}(s) = \mathbf{M}^T\mathbf{H}(s) + \mathbf{H}(s)\mathbf{M}, \tag{7.8}$$

where $\mathbf{M} \equiv \dot{\mathbf{E}}(0)$ and where the dot denotes differentiation with respect to s. The unique solution such that $\mathbf{H}(0) = \mathbf{1}$ is easily verified to be

$$\mathbf{H}(s) = e^{-s\mathbf{M}^T}e^{-s\mathbf{M}}. \tag{7.9}$$

Since histories are defined only when $s \geqq 0$, this result has been derived only for that range. Nevertheless, the difference equation (7.7) serves to define $\mathbf{H}(s)$ for negative s as well and shows that $\mathbf{H}(s)$ is analytic. By the principle of analytic continuation, (7.9), being analytic, is the unique solution for all s, when $\mathbf{E}(t)$ is given. If we substitute (7.9) back into (7.7), we obtain

$$\mathbf{E}(t)e^{t\mathbf{M}}[\mathbf{E}(t)e^{t\mathbf{M}}]^T = \mathbf{1}. \tag{7.10}$$

Hence $\mathbf{E}(t)e^{t\mathbf{M}}$ is an orthogonal tensor, say $\overline{\mathbf{Q}}(t)$. By (7.6) then

$$\mathbf{F}_0(t) = \mathbf{Q}(t)\overline{\mathbf{Q}}(t)e^{-t\mathbf{M}}. \tag{7.11}$$

The form asserted by NOLL's theorem holds trivially if $\mathbf{M} = \mathbf{0}$; if $\mathbf{M} \neq \mathbf{0}$, it follows if we set

$$\kappa \equiv |\mathbf{M}|, \qquad \mathbf{N}_0 \equiv \frac{1}{\kappa}\mathbf{M}. \tag{7.12}$$

(The proof reveals that the tensor $\mathbf{Q}(t)$ occurring in the result (7.3) is not necessarily the same orthogonal tensor as that occurring in the hypothesis (7.1).) Conversely, if (7.3) holds, an easy calculation shows that the motion is one of constant stretch history.

Exercise 7.1. Prove that in a motion with constant stretch history

$$F_t(\tau) = Q(\tau)Q(t)^T e^{(\tau - t)G}, \tag{7.13}$$

where

$$G = \kappa N \equiv \kappa Q(t)N_0 Q(t)^T, \qquad |N| = 1; \tag{7.14}$$

also

$$C_t^t(s) = e^{-sG^T} e^{-sG},$$

$$A_1 = -\dot{C}_t^t(0) = G^T + G = \kappa(N + N^T),$$

$$A_2 = \ddot{C}_t^t(0) = G^T A_1 + A_1 G = \kappa^2(2N^T N + N^2 + N^{T^2}), \tag{7.15}$$

$$A_3 = G^T A_2 + A_2 G,$$

$$\vdots \qquad \vdots$$

$$A_k = G^T A_{k-1} + A_{k-1} G.$$

A motion with constant stretch history is isochoric if and only if

$$\mathrm{tr} N_0 = 0 \tag{7.16}$$

Determination of the Deformation History from the First Three Rivlin-Ericksen Tensors.

With the aid of these results we are able to see easily the extremely special nature of motions with constant stretch history, which is expressed by WANG's corollary: The relative deformation history $C_t^t(s)$ of a motion with constant stretch history is determined uniquely by its first three Rivlin-Ericksen tesnors. That is, if three tensors $A_1(t)$, $A_2(t)$, and $A_3(t)$ are given, they can be the first three Rivlin-Ericksen tensors corresponding to at most one constant relative stretch history C_t^t.

The proof rests upon a simple lemma. Let $\tilde{S}$ be a symmetric tensor and $\tilde{W}$ a skew tensor in 3-dimensional space. Without loss of generality we can take the matrices of these tensors as having the forms

$$[\tilde{S}] = \begin{Vmatrix} a & 0 & 0 \\ 0 & b & 0 \\ 0 & 0 & c \end{Vmatrix}, \qquad [\tilde{W}] = \begin{Vmatrix} 0 & x & y \\ -x & 0 & z \\ -y & -z & 0 \end{Vmatrix}. \tag{7.17}$$

Then

$$[\tilde{S}\tilde{W} - \tilde{W}\tilde{S}] = \begin{Vmatrix} 0 & (a-b)x & (a-c)y \\ (a-b)x & 0 & (b-c)z \\ (a-c)y & (b-c)z & 0 \end{Vmatrix}. \tag{7.18}$$

Hence $\tilde{S}$ and $\tilde{W}$ commute if and only if

$$(a - b)x = 0, \quad (a - c)y = 0, \quad (b - c)z = 0. \tag{7.19}$$

Consequently, if $\tilde{S}$ has distinct proper numbers, it commutes with no other skew tensor than $\tilde{0}$. If $a = b \neq c$, $\tilde{S}$ commutes with $\tilde{W}$ if and only if $y = z = 0$. If $a = b = c$, $\tilde{S}$ commutes with all $\tilde{W}$.

WANG's corollary may now be proved in stages. Assume first that the proper numbers of $\tilde{A}_1$ are distinct. If two constant relative stretch histories $\tilde{C}_t^t$ can correspond to $\tilde{A}_1$ and $\tilde{A}_2$, then there exist tensors $\tilde{G}$ and $\bar{\tilde{G}}$ such that, by $(7.15)_{3,6}$,

$$\begin{aligned} \tilde{G} + \tilde{G}^T &= \bar{\tilde{G}} + \bar{\tilde{G}}^T, \\ \tilde{G}^T\tilde{A}_1 + \tilde{A}_1\tilde{G} &= \bar{\tilde{G}}^T\tilde{A}_1 + \tilde{A}_1\bar{\tilde{G}}. \end{aligned} \tag{7.20}$$

The first of these equations asserts that $\tilde{G} - \bar{\tilde{G}}$ is skew; the second, that $\tilde{G} - \bar{\tilde{G}}$ commutes with $\tilde{A}_1$. By the lemma, $\tilde{G} - \bar{\tilde{G}} = \tilde{0}$.

Suppose now $\tilde{A}_1$ has two and only two distinct proper numbers. Then relative to a suitable orthonormal basis

$$[\tilde{A}_1] = \begin{Vmatrix} a & 0 & 0 \\ 0 & a & 0 \\ 0 & 0 & b \end{Vmatrix}, \quad a \neq b. \tag{7.21}$$

Case 1. Assume that, relative to this same basis,

$$[\underset{\sim}{A}_2] = \begin{Vmatrix} u & 0 & 0 \\ 0 & u & 0 \\ 0 & 0 & v \end{Vmatrix}. \tag{7.22}$$

The most general $\underset{\sim}{G}$ compatible with $(7.15)_3$ and (7.21) is given by

$$[\underset{\sim}{G}] = \begin{Vmatrix} \frac{1}{2}a & x & y \\ -x & \frac{1}{2}a & z \\ -y & -z & \frac{1}{2}b \end{Vmatrix}. \tag{7.23}$$

By (7.22)

$$[\underset{\sim}{G}^T\underset{\sim}{A}_1 + \underset{\sim}{A}_1\underset{\sim}{G}] = \begin{Vmatrix} a^2 & 0 & (a-b)y \\ 0 & a^2 & (a-b)z \\ (a-b)y & (a-b)z & b^2 \end{Vmatrix}. \tag{7.24}$$

Since $a \neq b$, it follows from $(7.15)_6$ and (7.22) that

$$u = a^2, \quad v = b^2, \quad y = 0, \quad z = 0. \tag{7.25}$$

Exercise 7.2. Using (7.23), (7.25), and $(7.15)_1$ and the lemma, show that in fact

$$\underset{\sim}{C}_t^t(s) = e^{-s\underset{\sim}{A}_1}. \tag{7.26}$$

Case 2. Assume that (7.22) does not hold, and that two constant relative stretch histories can correspond to $\underset{\sim}{A}_1$, $\underset{\sim}{A}_2$, and $\underset{\sim}{A}_3$. Then, again, there exist $\underset{\sim}{G}$ and $\bar{\underset{\sim}{G}}$ such that (7.21) holds. Since $\underset{\sim}{G} - \bar{\underset{\sim}{G}}$ is a skew tensor that commutes with $\underset{\sim}{A}_1$ as given by (7.21), the lemma shows that

$$[\underset{\sim}{G} - \bar{\underset{\sim}{G}}] = \begin{Vmatrix} 0 & x & 0 \\ -x & 0 & 0 \\ 0 & 0 & 0 \end{Vmatrix}. \tag{7.27}$$

But also, by $(7.15)_8$

$$\tilde{G}^T\tilde{A}_2 + \tilde{A}_2\tilde{G} = \bar{\tilde{G}}^T\tilde{A}_2 + \tilde{A}_2\bar{\tilde{G}}, \tag{7.28}$$

so that $\tilde{G} - \bar{\tilde{G}}$ commutes with $\tilde{A}_2$.

Exercise 7.3. Recalling that (7.22) does not hold, prove that $\tilde{G} = \bar{\tilde{G}}$ in Case 2. Finally, if $\tilde{A}_1 = \alpha\tilde{1}$, prove that (7.26) holds, and hence complete the argument for WANG's corollary.

Accordingly, then, three given tensors $\tilde{A}_1(t)$, $\tilde{A}_2(t)$, and $\tilde{A}_3(t)$ can be the Rivlin-Ericksen tensors corresponding to at most one $\tilde{C}_t^t(s)$ belonging to a motion with constant stretch history. In general, three tensors taken arbitrarily will fail to belong to any motion with constant stretch history.

While NOLL's theorem clearly is independent of the dimension of the space, WANG's corollary rests heavily on use of the dimension 3.

Equivalence of Simple Materials with Rivlin-Ericksen Materials in Motions with Constant Stretch History.

In view of WANG's corollary, any information that can be determined in a motion with constant stretch history from $\tilde{C}_t^t(s)$ can be determined also from $\tilde{A}_1(t)$, $\tilde{A}_2(t)$, $\tilde{A}_3(t)$. Therefore any functional of $\tilde{C}_t^t(s)$, equals, in these motions, a function of $\tilde{A}_1(t)$, $\tilde{A}_2(t)$, $\tilde{A}_3(t)$. Consequently the general constitutive equation (4.7) may be replaced as far as motions with constant stretch history are concerned, by

$$\bar{\tilde{T}}(t) = \tilde{\mathfrak{f}}\,[\bar{\tilde{A}}_1(t), \bar{\tilde{A}}_2(t), \bar{\tilde{A}}_3(t), \tilde{C}(t)], \tag{7.29}$$

where $\bar{\tilde{A}}_k(t) \equiv \tilde{R}(t)^T\tilde{A}_k(t)\tilde{R}(t)$, and where $\tilde{\mathfrak{f}}$ is a function. A material which satisfies (7.29) for all motions is called a Rivlin-Ericksen material of complexity 3. According to (7.29), then, in the class of motions with constant stretch history a simple material cannot be distinguished from the far more special Rivlin-Ericksen material of complexity 3. For an isotropic material, (7.29) becomes

$$\tilde{T}(t) = \tilde{\mathfrak{f}}\,[\tilde{A}_1(t), \tilde{A}_2(t), \tilde{A}_3(t), B(t)], \tag{7.30}$$

and for a fluid

$$\underset{\sim}{T}(t) = -p\underset{\sim}{1} + \underset{\sim}{\mathfrak{f}}(\underset{\sim}{A}_1(t), \underset{\sim}{A}_2(t), \underset{\sim}{A}_3(t), \rho), \tag{7.31}$$

where the functions $\underset{\sim}{\mathfrak{f}}$, in the two cases, are isotropic in the sense that for all symmetric $\underset{\sim}{A}_1$, $\underset{\sim}{A}_2$, $\underset{\sim}{A}_3$, $\underset{\sim}{B}$, and all orthogonal $\underset{\sim}{Q}$

$$\begin{aligned} &\underset{\sim}{\mathfrak{f}}(\underset{\sim}{Q}\underset{\sim}{A}_1\underset{\sim}{A}^T, \underset{\sim}{Q}\underset{\sim}{A}_2\underset{\sim}{Q}^T, \underset{\sim}{Q}\underset{\sim}{A}_3\underset{\sim}{Q}^T, \underset{\sim}{Q}\underset{\sim}{B}\underset{\sim}{Q}^T \text{ or } \rho), \\ &= \underset{\sim}{Q}\,\underset{\sim}{\mathfrak{f}}(\underset{\sim}{A}_1, \underset{\sim}{A}_2, \underset{\sim}{A}_3, \underset{\sim}{B} \text{ or } \rho)\underset{\sim}{Q}^T. \end{aligned} \tag{7.32}$$

These results, which express <u>WANG's theorem</u>, may be interpreted in two ways. On the one hand, they enable us to solve easily special problems concerned with motions of constant stretch history. However complicated may be in general the response of a material, in these particular motions we need consider only a simple special constitutive equation. On the other hand, they show that observation of this class of flows is insufficient to tell us much about a material, since most of the complexities of material response are prevented from manifesting themselves.

<u>Classification of Motions with Constant Stretch History.</u>

Returning to NOLL's theorem (7.3), we see at once that it suggest an invariant classification of all motions with constant stretch history into three mutually exclusive types:

1. $\underset{\sim}{N}_0^2 = \underset{\sim}{0}$. These motions are called <u>viscometric flows</u>.
2. $\underset{\sim}{N}_0^3 = \underset{\sim}{0}$ but $\underset{\sim}{N}_0^2 \neq \underset{\sim}{0}$.
3. $\underset{\sim}{N}_0$ is not nilpotent

In types 1 and 2, since $\mathrm{tr}\underset{\sim}{N}_0 = 0$, the motion is isochoric.

There are interesting examples of all three types, but the simplest, the viscometric flows, are used most in applications.

PART II. FLUIDS

LECTURE 8. THE STRESS SYSTEM IN VISCOMETRIC FLOWS OF INCOMPRESSIBLE FLUIDS

Recapitulation.

From the theorems of NOLL and WANG, stated and proved in the last lecture, it follows that in a viscometric flow of an incompressible fluid,

$$\tilde{T} = -p\tilde{1} + \tilde{\mathcal{G}}(\tilde{A}_1, \tilde{A}_2), \tag{8.1}$$

where

$$\begin{aligned} \tilde{A}_1 &= \kappa(\tilde{N} + \tilde{N}^T), \\ \tilde{A}_2 &= 2\kappa^2\tilde{N}^T\tilde{N}, \\ \tilde{A}_n &= \tilde{0}, \text{ if } n \geqq 3, \end{aligned} \tag{8.2}$$

and

$$\tilde{N}^2 = \tilde{0}, \quad \mathrm{tr}\tilde{N} = 0, \quad |\tilde{N}| = 1, \tag{8.3}$$

and

$$\tilde{\mathcal{G}}(\tilde{Q}\tilde{A}_1\tilde{Q}^T; \tilde{Q}\tilde{A}_2\tilde{Q}^T) = \tilde{Q}\,\tilde{\mathcal{G}}(\tilde{A}_1, \tilde{A}_2)\tilde{Q}^T$$

for all symmetric tensors $\tilde{A}_1$ and $\tilde{A}_2$ and for all orthogonal tensors $\tilde{Q}$.

Following the method of COLEMAN and NOLL, we shall now determine the most general stress system compatible with the equations just stated.

Functional Equations for the Response Function.

First of all, (8.1) and (8.2) show that

$$\mathbf{T} = -p\mathbf{1} + \mathfrak{f}(\kappa, \mathbf{N}). \tag{8.5}$$

Second, if we replace κ by $\alpha\kappa$, where $\alpha = \pm 1$, and $\mathbf{N}$ by $\frac{1}{\alpha}\mathbf{Q}\mathbf{N}\mathbf{Q}^T$, by (8.2) $\mathbf{A}_1$ is replaced by $\mathbf{Q}\mathbf{A}_1\mathbf{Q}^T$, and $\mathbf{A}_2$ is replaced by $\mathbf{Q}\mathbf{A}_2\mathbf{Q}^T$. Therefore, (8.4) is equivalent to

$$\mathfrak{f}\left(\alpha\kappa, \frac{1}{\alpha}\mathbf{Q}\mathbf{N}\mathbf{Q}^T\right) = \mathbf{Q}\,\mathfrak{f}(\kappa, \mathbf{N})\mathbf{Q}^T, \tag{8.6}$$

where $\mathfrak{f}$ is the function occurring in (8.5). In view of (8.1) we may describe as follows the invariance asserted by the functional equation (8.6): To replace κ by $\alpha\kappa$ and N by $\frac{1}{\alpha}\mathbf{Q}\mathbf{N}\mathbf{Q}^T$, where $\alpha = \pm 1$ and Q is any orthogonal tensor, results in replacing $\mathbf{T}$ by $\mathbf{Q}\mathbf{T}\mathbf{Q}^T$.

In view of (8.3), we may choose a basis such that

$$[\mathbf{N}] = \begin{Vmatrix} 0 & 0 & 0 \\ 1 & 0 & 0 \\ 0 & 0 & 0 \end{Vmatrix}. \tag{8.7}$$

By (7.14), the basis with respect to which $[\mathbf{N}]$ has this special form is generally a rotating one, and it need not be the natural basis of any co-ordinate system.

Illustration by means of Shearing Flow.

While the results we shall deduce follow from the algebraic formulae just given, it is easier to visualize them in terms of a special case. To this end we consider a shearing flow, given in a suitable rectangular Cartesian system by the velocity components

$$\dot{x}_1 = 0, \quad \dot{x}_2 = v(x_1), \quad \dot{x}_3 = 0. \tag{8.8}$$

Exercise 8.1. Show from (1.37) and (1.39) that for the flow (8.8)

$$\mathbf{F}_t(\tau) = \mathbf{1} + (\tau - t)\kappa\mathbf{N} = e^{(\tau - t)\kappa\mathbf{N}}, \tag{8.9}$$

where $\mathbf{N}$ has the form (8.7) with respect to the co-ordinate basis and where

$$\kappa = v'(x_1). \tag{8.10}$$

In this example of a viscometric flow, the particles move in straight lines at uniform speed, and κ is twice the only non-vanishing principal stretching. It is customary to refer to κ as the shearing, not only in the special case (8.10) but also for any viscometric flow.

Consequences of Invariance under Reflections.

We shall now determine the most general stress system compatible with the constitutive equation (8.1) in a viscometric flow, characterized by (8.2). We shall motivate the results by describing them in terms of the special case (8.7).

First, a reflection in the plane normal to the flow should be expected to leave the whole stress system invariant. To see if it does, take $\tilde{Q}$ such that

$$[\tilde{Q}] = \begin{Vmatrix} 1 & 0 & 0 \\ 0 & 1 & 0 \\ 0 & 0 & -1 \end{Vmatrix}, \tag{8.11}$$

and take $\alpha = 1$. From (8.6), $\tilde{Q}\tilde{N}\tilde{Q}^T = \tilde{N}$. The assertion embodied in (8.5) is that this same transformation carries $\tilde{T}$ into $\tilde{Q}\tilde{T}\tilde{Q}^T$. But by (8.11)

$$[\tilde{Q}\tilde{T}\tilde{Q}^T] = \begin{Vmatrix} T\langle 11\rangle & T\langle 12\rangle & -T\langle 13\rangle \\ \cdot & T\langle 22\rangle & -T\langle 23\rangle \\ \cdot & \cdot & T\langle 33\rangle \end{Vmatrix}, \tag{8.12}$$

where the $T\langle ij\rangle$ are the components of $\tilde{T}$ with respect to the basis in which $\tilde{N}$ and $\tilde{Q}$ have the forms (8.6) and (8.11). Thus in order that $\tilde{Q}\tilde{T}\tilde{Q}^T = \tilde{T}$, it is necessary that

$$T\langle 13\rangle = 0, \qquad T\langle 23\rangle = 0. \tag{8.13}$$

By (8.4), the remaining components of $\tilde{T} + p\tilde{1}$ with respect to this basis are functions of κ only:

$$\begin{aligned} T\langle 12\rangle &= \tau(\kappa), \\ T\langle 11\rangle - T\langle 33\rangle &= \sigma_1(\kappa), \\ T\langle 22\rangle - T\langle 33\rangle &= \sigma_2(\kappa). \end{aligned} \tag{8.14}$$

If we prefer an invariant form, without mention of a basis, it may be written down by combining (8.14), (8.5), and (8.6):

$$\underset{\sim}{T} = -p\underset{\sim}{1} + \tau(\kappa)(\underset{\sim}{N} + \underset{\sim}{N}^T) + \sigma_1(\kappa)\underset{\sim}{N}^T\underset{\sim}{N} + \sigma_2(\kappa)\underset{\sim}{N}\underset{\sim}{N}^T, \tag{8.15}$$

where p is not determined by the constitutive equation. (In general, $p = -T<33> \neq -\frac{1}{3}\,\mathrm{tr}\,\underset{\sim}{T}$.)

Now we consider a reflection in the direction of flow. We expect that such a reflection, which amounts to replacing κ by $-\kappa$, should reverse the shear stress $T<12>$ but leave all normal tractions $T<11>$, $T<22>$, $T<33>$ unchanged. Accordingly, we take $\alpha = -1$ and $\underset{\sim}{Q}$ such that

$$[\underset{\sim}{Q}] = \left\| \begin{matrix} 1 & 0 & 0 \\ 0 & -1 & 0 \\ 0 & 0 & 1 \end{matrix} \right\|. \tag{8.16}$$

Then, by (8.6)

$$\frac{1}{\alpha}\,\underset{\sim}{Q}\underset{\sim}{N}\underset{\sim}{Q}^T = -\underset{\sim}{Q}\underset{\sim}{N}\underset{\sim}{Q}^T = \underset{\sim}{N}. \tag{8.17}$$

Under this transformation, κ goes into $-\kappa$ and $\underset{\sim}{T}$ goes into $\underset{\sim}{Q}\underset{\sim}{T}\underset{\sim}{Q}^T$. By (8.16),

$$[\underset{\sim}{Q}\underset{\sim}{T}\underset{\sim}{Q}^T] = \left\| \begin{matrix} T<11> & -T<12> & 0 \\ \cdot & T<22> & 0 \\ \cdot & \cdot & T<33> \end{matrix} \right\| \tag{8.18}$$

Thus change of the sign of κ changes $T<12>$ into $-T<12>$ but leaves the remaining stresses unaffected. By (8.14), then,

$$\begin{aligned} \tau(-\kappa) &= -\tau(\kappa), \\ \sigma_1(-\kappa) &= \sigma_1(\kappa), \\ \sigma_2(-\kappa) &= \sigma_2(\kappa). \end{aligned} \tag{8.19}$$

That is, τ is an odd function of the shearing, while σ_1 and σ_2 are even functions. From (8.19) it follows that

$$\tau(0) = 0. \tag{8.20}$$

If, as is customary, p is normalized so that $\mathfrak{g}(\mathbf{0}, \mathbf{0}) = \mathbf{0}$, then also

$$\begin{aligned} \sigma_1(0) &= 0, \\ \sigma_2(0) &= 0. \end{aligned} \tag{8.21}$$

<u>Exercise 8.2</u>. Prove that if (8.15) and (8.21) hold, then (8.4) and (8.5) are satisfied.

<u>The Viscometric Functions. Normal-Stress Effects.</u>

What has been shown, then, is that the stress system in any incompressible simple fluid in a viscometric flow is given by (8.15), with the functions τ, σ_1, and σ_2 restricted by (8.19) and (8.21).

The functions τ, σ_1, and σ_2 are the <u>viscometric functions</u> of the simple fluid whose constitutive equation reduces to (8.15) in a viscometric flow. Obviously, infinitely many different fluids share the same set of three viscometric functions.

In the particular case when the velocity field is given by (8.8), the basis with respect to which $\mathbf{N}$ has the form (8.7) is the natural basis of the fixed Cartesian co-ordinate system, and the component relations (8.14) may be interpreted in terms of Cartesian co-ordinates. In the further special case when

$$v(x_1) = \kappa x_1, \qquad \kappa = \text{const.} \tag{8.22}$$

the motion (8.8) reduces to the simple shearing (5.11). This motion has already been shown to be possible, subject to boundary tractions alone, in any homogeneous simple material, compressible, or incompressible. <u>A fortiori</u>, it may be produced without application of body force in any homogeneous incompressible fluid. In this case, the stress system (8.14) may be interpreted immediately. $\tau(\kappa)$ gives the shear stress that must be supplied on a plane x_1 = const. in order to produce the flow. The normal traction on this same plane is $T_{\langle 11\rangle}$, and that on the flow plane is $T_{\langle 33\rangle}$. In view of the term $-p\mathbf{1}$ in (8.1), either of these, but not both, may be given any value, at pleasure. If we choose to leave the planes x_3 = const. free, then $T<33> = 0$, and (8.14) yields

$$\begin{aligned} T<11> &= \sigma_1(\kappa), \\ T<22> &= \sigma_2(\kappa). \end{aligned} \qquad (8.23)$$

Thus a fixed normal traction on the plane x_1 = const., determined by κ and by the nature of the fluid, must be supplied, and likewise a normal traction in the planes x_2 = const., which are normal to the flow. The necessity for these normal tractions is an example of what are called <u>normal-stress effects</u>. In particular, the result (8.23) shows that shear stress alone is insufficient to produce simple shearing. In addition, suitable and generally unequal normal tractions, determined by the nature of the fluid, are necessary in order for the flow to occur.

<u>Position of the Classical Theory</u>.

According to the classical or <u>Navier-Stokes theory</u> of fluids,

$$\begin{aligned} \tau(\kappa) &\equiv \mu\kappa, \text{ where } \mu = \text{const.}, \\ \sigma_1(\kappa) &\equiv \sigma_2(\kappa) \equiv 0. \end{aligned} \qquad (8.24)$$

If we assume that τ and σ_1 and σ_2 have three continuous derivatives at $\kappa = 0$, then, by (8.19) and (8.21),

$$\begin{aligned} \tau(\kappa) &= \mu\kappa + \mu'\kappa^3 + 0(\kappa^5), \\ \sigma_1(\kappa) &= s_1\kappa^2 + 0(\kappa^4), \\ \sigma_2(\kappa) &= s_2\kappa^2 + 0(\kappa^4), \end{aligned} \qquad (8.25)$$

where μ, μ', s_1, and s_2 are constants.

Thus the effects of second-order in κ are normal-stress effects, while departure from the classical proportionality of shear stress to shearing is an effect of third order in κ. Roughly speaking, departures from the classical behavior (8.24) may be expected to be observed for σ_1 and σ_2 at lower shearings κ than for τ. Still more roughly, normal-stress effects can be expected to manifest themselves within the range in which the response of the shear stress remains classical.

In the example used to interpret the results so far, $v(x_1)$ in (8.8) has been assumed linear. This velocity field is identical with the one that follows in the same circumstances according to the Navier-Stokes theory, but the stress-system required in order to produce it has been shown to be different. Nevertheless, the flow is a possible one, subject to boundary tractions alone.

When, on the other hand, the velocity profile $v(x_1)$ in (8.8) fails to be linear, or when, as is the case in general, the basis with respect to which $\underset{\sim}{N}$ has the form (8.7) is a rotating one, we have no reason to expect a viscometric flow to be possible unless suitable non-conservative body forces be supplied. In the next lecture we shall consider some particular cases in which the dynamical equations can indeed be satisfied when $\underset{\sim}{b} = \underset{\sim}{0}$. As will be seen, the assignment of speeds to the stream-lines will generally be entirely different from that required by the Navier-Stokes theory.

LECTURE 9: DYNAMICAL CONDITIONS IN VISCOMETRIC FLOWS.

Recapitulation.

The stress corresponding to any viscometric flow of an incompressible simple fluid is given by (8.15):

$$\underset{\sim}{S} \equiv \underset{\sim}{T} + p\underset{\sim}{1} = \tau(\kappa)(\underset{\sim}{N} + \underset{\sim}{N}^T) + \sigma_1(\kappa)\underset{\sim}{N}^T\underset{\sim}{N} + \sigma_2(\kappa)\underset{\sim}{N}\underset{\sim}{N}^T . \qquad (8.15)$$

The viscometric functions $\tau(\kappa)$, $\sigma_1(\kappa)$, $\sigma_2(\kappa)$ are determined uniquely by the constitutive equation and hence are the same for all viscometric flows of any given fluid. κ and $\underset{\sim}{N}$ generally are functions of both place and time. κ is a scalar, which may be called the <u>shearing</u>, and $\underset{\sim}{N}$ is a tensor such that $|\underset{\sim}{N}| = 1$, $\underset{\sim}{N}^2 = \underset{\sim}{0}$ (and hence $\mathrm{tr}\underset{\sim}{N} = 0$). The orthonormal basis with respect to which $\underset{\sim}{N}$ has the special component matrix (8.7) may vary with place and time and need not be the natural basis of any co-ordinate system. The scalar p, which equals $-T\langle 33\rangle$ in the special basis, is not determined by the deformation history. In general, (8.15) will fail to satisfy Cauchy's law (3.16) unless a suitable body force be supplied.

Dynamical Compatibility.

Of greatest interest are such particular viscometric flows as may be effected by applying suitable boundary tractions alone. One such has already been exhibited, the rectilinear shearing, defined by (8.8):

$$\dot{x}_1 = 0 , \qquad \dot{x}_2 = v(x_1) \qquad \dot{x}_3 = 0 , \qquad (8.8)$$

in the special case when $v(x_1) = \kappa x_1$, $\kappa = \mathrm{const}$. This case is not typical in that one and the same velocity field is possible in all fluids. Generally the velocity field that meets given dynamical requirements depends upon the viscometric functions, as we shall see.

Shearing Flow.

We shall now find the most general shearing flow (8.8) that can be effected by boundary tractions and conservative body force in a simple fluid whose viscometric functions are $\tau(\kappa)$, $\sigma_1(\kappa)$, and $\sigma_2(\kappa)$. For shearing flow, the basis with respect to which $\mathbf{N}$ has the special form (8.7) is the natural basis of the co-ordinate system used, and $\mathbf{N}$ = const. As has been shown already,

$$\kappa = v'(x_1) \quad . \tag{8.10}$$

We assume that ρ = const.

We shall employ Cauchy's first law in the form (5.17):

$$\operatorname{div} \mathbf{S} - \rho \operatorname{grad} \phi = \rho \ddot{\mathbf{x}} \ . \tag{5.17}$$

By (8.15) and (8.8), $\mathbf{S}$ is a function of x_1 only, and $\ddot{\mathbf{x}} = \mathbf{0}$. Hence (5.17) is equivalent to the following three differential equations:

$$\partial_{x_1} S\langle 11\rangle - \rho \partial_{x_1} \phi = 0 \ ,$$

$$\partial_{x_1} S\langle 12\rangle - \rho \partial_{x_2} \phi = 0 \ , \tag{9.1}$$

$$\rho \partial_{x_3} \phi = 0 \ .$$

The last of these equations shows that ϕ is independent of x_3. From the first two, since $\mathbf{S}$ is a function of x_1 only,

$$\partial^2_{x_2} \phi = 0 \quad , \qquad \partial_{x_1} \partial_{x_2} \phi = 0 \ . \tag{9.2}$$

Hence ϕ is linear in x_2:

$$\rho\phi = - a x_2 + k(x_1) + h(t) \ , \tag{9.3}$$

where a is a constant, called the <u>specific driving force</u> in this flow, and where the functions $k(x_1)$ and $h(t)$ are arbitrary. Substituting (9.3) into $(9.1)_2$ and integrating yields

$$T\langle 12\rangle = S\langle 12\rangle = - a x_1 + c \ , \tag{9.4}$$

where c = const:

No matter what be the velocity profile $v(x_1)$, the shear stress must be a linear function of x_1. Moreover, the shear stress is the same in all fluids; it is determined by the driving force and is independent of the viscometric functions. Substitution of (9.3) into $(9.1)_1$ and integrating yields

$$S\langle 11\rangle = k(x_1) + b \ , \tag{9.5}$$

where $b = \text{const.}$ Conversely, if (9.3), (9.4), and (9.5) hold, the dynamical equations (9.1) are satisfied.

The entire stress system may be calculated as follows. First, by (9.5) and (9.3)

$$\begin{aligned} T\langle 11\rangle &= S\langle 11\rangle - p = S\langle 11\rangle - (p + \rho\upsilon) + \rho\upsilon \\ &= ax_2 + b + \rho\upsilon - h(t) \ . \end{aligned} \tag{9.6}$$

Since $\partial_{x_2}[T\langle 11\rangle - \rho\upsilon] = a$, the constant a is the gradient of $T\langle 11\rangle - \rho\upsilon$ in the direction of flow. Second, by this result and (8.14),

$$\begin{aligned} T\langle 22\rangle &= (T\langle 22\rangle - T\langle 11\rangle) + T\langle 11\rangle \ , \\ &= \sigma_2(\kappa) - \sigma_1(\kappa) + ax_2 + b + \rho\upsilon - h(t) \ , \\ T\langle 33\rangle &= (T\langle 33\rangle - T\langle 11\rangle) + T\langle 11\rangle \ , \\ &= -\sigma_1(\kappa) + ax_2 + b + \rho\upsilon - h(t) \ . \end{aligned} \tag{9.7}$$

With $v(x_1)$ arbitrary and κ given by (8.10), the normal stresses are delivered by these formulae. Combining (9.4) and $(8.14)_1$ yields a differential equation to determine $v(x_1)$:

$$\tau(\kappa) = \tau\,[v'(x_1)] = -ax_1 + c \ . \tag{9.8}$$

The arbitrary constants a and c are to be assigned, and then $v(x_1)$ is determined by integrating (9.8).

If we take $a = 0$ and assume that κ is invertible, (9.8) requires that $\kappa = \text{const.}$, and we recover the results already obtained for simple shearing. The foregoing analysis shows simple shearing to be the only rectilinear shearing flow (8.8) that can be produced without specific driving force, provided the body force be conservative.

Channel Flow.

We shall now seek a solution that represents the flow of a mass of material adhering to stationary infinite plates $x_1 = \pm d$. Thus we require $v(x_1)$ to be such that

$$v(d) = v(-d) = 0 . \tag{9.9}$$

We assume that the shear viscosity function $\tau(\kappa)$ is invertible with inverse, say, ζ. Any inverse is necessarily odd. Then (9.8) yields

$$\kappa = v'(x_1) = \zeta(-ax_1 + c) . \tag{9.10}$$

Exercise 9.1. Prove that, since ζ is an odd function, (9.10) is compatible with (9.9) if and only if $c = 0$.

Consequently

$$v(x_1) = \int_{x_1}^{d} \zeta(a\xi)d\xi . \tag{9.11}$$

The velocity profile $v(x_1)$ is thus determined uniquely by the shear-viscosity function $\tau(\kappa)$. In contrast to the case of simple shearing, however, the profile is generally not at all the same as that predicted by the Navier-Stokes equations. Indeed, if $(8.24)_1$ holds, then $\zeta(\xi) = (1/\mu)\xi$, and (9.11) yields

$$v(x_1) = \frac{1}{\mu}\int_{x_1}^{d} a\xi d\xi = \frac{a}{2\mu}(d^2 - x_1^2) , \tag{9.12}$$

the classical parabolic form. Conversely, if (9.12) holds, $\zeta(a\xi)$ is a linear function of a, and the classical linear formula (8.24) for shear viscosity results.

The discharge D, the volume of fluid passing through unit depth of channel in unit time, is given by

$$\begin{aligned} D \equiv \int_{-d}^{d} v(x)dx &= 2\int_{0}^{d} dx \int_{x}^{d} \zeta(a\xi)d\xi , \\ &= \frac{2}{a^2}\int_{0}^{ad} \xi\zeta(\xi)d\xi . \end{aligned} \tag{9.13}$$

Conversely, if the discharge $D(a,d)$ is known as a function of a and d, (9.13) yields

$$\tau^{-1}(ad) = \zeta(ad) = \frac{1}{2ad^2}\,\partial_a[a^2D(a,d)] \;.$$

Thus $D(a,d)$ determines the shear-viscosity function $\tau(\xi)$ uniquely. In particular, the classical formula

$$D = \frac{2ad^3}{3\mu} \tag{9.15}$$

holds if and only if the shear-viscosity function is linear.

Velocity profile, discharge, and shear-viscosity function determine one another and are unaffected by the normal-stress functions σ_1 and σ_2. Thus if (9.15) holds, there is no reason at all to expect the remaining classical formulae $(8.24)_{3,4}$ to hold. Therefore, the classical viscometric tests, which refer to shear viscosity alone, do not tell much about the fluid being tested. If in a particular case a classical formula such as (9.15) emerges, this fact not only fails to show that the fluid tested obeys the Navier-Stokes equations but even is insufficient to establish the Navier-Stokes theory of viscometry. Additional measurements are necessary. In the present case, by (9.6), the normal tractions on the fixed plates do not differ from those predicted by the classical theory. By (9.7), however, those on the flow planes (x_3 = const.) and those on planes normal to the flow (x_2 = const.) may be entirely different.

Since these normal tractions are difficult to interpret, we turn to a different class of flows, in which normal-stress effects are more striking.

Helical Flows.

Consider the velocity field whose contravariant components in a cylindrical polar co-ordinate system r, θ, z are given by

$$\dot{r} = 0 \;, \quad \dot{\theta} = \omega(r) \;, \quad \dot{z} = u(r). \tag{9.16}$$

Each particle remains upon a fixed cylinder r = const., on which it describes a helix, whose pitch is the same for all the particles on any one cylinder. We set

$$f(r) \equiv \omega'(r) \;, \quad h(r) \equiv u'(r) \;. \tag{9.17}$$

<u>Exercise 9.2</u>. Prove that a helical flow is a flow with constant stretch history and that

$$\kappa = \sqrt{[r^2 f(r)^2 + h(r)^2]} \quad . \tag{9.18}$$

Let $\mathbf{e}_i(\mathbf{x})$, $i = 1, 2, 3$, be an orthonormal basis tangent to the co-ordinate curves at $\mathbf{x}$, and let

$$\mathbf{i}_1 = \mathbf{e}_1 \ , \quad \mathbf{i}_2 = \alpha\mathbf{e}_2 + \beta\mathbf{e}_3 \ , \quad \mathbf{i}_3 = -\beta\mathbf{e}_2 + \alpha\mathbf{e}_3 \ , \tag{9.19}$$

where

$$\alpha = \frac{r}{\kappa} f(r) \ , \quad \beta = \frac{1}{\kappa} h(r) \ , \quad \alpha^2 + \beta^2 = 1 \ . \tag{9.20}$$

Prove that $\mathbf{N}_0$ has the form (8.7) with respect to this basis.

From the results of the foregoing problem, the formulae (8.14) are valid for the components of $\mathbf{T}$ relative to the basis $\mathbf{i}_1$, $\mathbf{i}_2$, $\mathbf{i}_3$:

$$\begin{aligned} &T<12> = \tau(\kappa) \ , \quad T<13> = 0 \ , \quad T<23> = 0 \ , \\ &T<11> - T<33> = \sigma_1(\kappa) \\ &T<22> - T<33> = \sigma_2(\kappa) \ . \end{aligned} \tag{9.21}$$

The physical components of $\mathbf{T}$ in cylindrical co-ordinates are its components with respect to the orthonormal basis $\mathbf{e}_i$. Denoting these components by $T<rr>$, $T<r\theta>$, <u>etc</u>., we find that

$$\begin{aligned} T<r\theta> &= \mathbf{e}_1 \cdot \mathbf{T}\mathbf{e}_2 = \mathbf{i}_1 \cdot (\alpha\mathbf{T}\mathbf{i}_2 - \beta\mathbf{T}\mathbf{i}_3) \\ &= \alpha T<12> - \beta T<13> \ , \end{aligned} \tag{9.22}$$

$$\begin{aligned} T<\theta z> &= \mathbf{e}_2 \cdot \mathbf{T}\mathbf{e}_3 = (\alpha\mathbf{i}_2 - \beta\mathbf{i}_3) \cdot (\beta\mathbf{T}\mathbf{i}_2 + \alpha\mathbf{T}\mathbf{i}_3) \ , \\ &= \alpha\beta(T<22> - T<33>) + (\alpha^2 - \beta^2)T<23> \ , \end{aligned}$$

<u>etc</u>. From these results and (9.21) we find the stress system in terms of the viscometric functions:

$$T\langle r\theta\rangle = \alpha\tau(\kappa) \;,$$

$$T\langle rz\rangle = \beta\tau(\kappa) \;,$$

$$T\langle \theta z\rangle = \alpha\beta\sigma_2(\kappa) \;, \tag{9.23}$$

$$T\langle rr\rangle - T\langle zz\rangle = \sigma_1(\kappa) - \beta^2\sigma_2(\kappa) \;,$$

$$T\langle \theta\theta\rangle - T\langle zz\rangle = (\alpha^2 - \beta^2)\,\sigma_2(\kappa) \;.$$

It remains now to see whether the functions $f(r)$ and $h(r)$ can be chosen in such a way as to make these stresses compatible with Cauchy's first law of motion.

<u>Exercise 9.3</u>. Since $\underset{\sim}{S}$ is a function of r only, prove that for a helical flow of an incompressible fluid Cauchy's first law assumes the form

$$\partial_r S\langle rr\rangle + \frac{1}{r}\,(S\langle rr\rangle - S\langle \theta\theta\rangle) - \rho\partial_r\phi = -\rho r\omega^2 \;,$$

$$r\partial_r S\langle r\theta\rangle + 2S\langle r\theta\rangle - \rho\partial_\theta\phi = 0 \;, \tag{9.24}$$

$$\partial_r S\langle rz\rangle + \frac{1}{r}\,S\langle rz\rangle - \rho\partial_z\phi = 0 \;.$$

Hence

$$\partial_r(r^2 T\langle r\theta\rangle) = -rd \;,$$

$$\partial_r(rT\langle rz\rangle) = -ra \;, \tag{9.25}$$

$$T\langle rr\rangle = \rho\upsilon + k(r,t) + az + d\theta \;,$$

$$\partial_r k(r,t) + \frac{1}{r}\,(T\langle rr\rangle - T\langle \theta\theta\rangle) = -\rho r\omega^2 \;,$$

where a and d are arbitrary constants.

Integration of the first two of these equations yields

$$T\langle r\theta\rangle = \left(\frac{c}{r^2} - \frac{d}{2}\right) \;, \tag{9.26}$$

$$T\langle rz\rangle = \left(\frac{b}{r} - \frac{ra}{2}\right) , \tag{9.26}$$

where b and c are arbitrary constants. These equations are compatible with $(9.23)_{1,2}$ if and only if

$$\tau(\kappa) = \gamma , \qquad \kappa = \zeta(\gamma) , \tag{9.27}$$

where

$$\gamma = \gamma(r) = \sqrt{\left[\left(\frac{c}{r^2} - \frac{d}{2}\right)^2 + \left(\frac{b}{r} - \frac{ra}{2}\right)^2\right]} . \tag{9.28}$$

By $(9.23)_{1,2}$, then,

$$\alpha = \frac{T\langle r\theta\rangle}{\tau(\kappa)} = \frac{1}{\gamma}\left(\frac{c}{r^2} - \frac{d}{2}\right) ,$$

$$\beta = \frac{T\langle rz\rangle}{\tau(\kappa)} = \frac{1}{\gamma}\left(\frac{b}{r} - \frac{ra}{2}\right) . \tag{9.29}$$

Finally, from (9.20) and (9.17) ,

$$\omega'(r) = f = \frac{\kappa\alpha}{r} = \frac{\zeta(\gamma)}{\gamma r}\left(\frac{c}{r^2} - \frac{d}{2}\right) ,$$

$$u'(r) = h = \kappa\beta = \frac{\zeta(\gamma)}{\gamma}\left(\frac{b}{r} - \frac{ra}{2}\right) . \tag{9.30}$$

When the four constants a, b, c, d are fixed, γ becomes a known function of r by (9.28). Hence the two functions $\omega(r)$ and $u(r)$ occurring in the velocity field (9.16) are determined to within six arbitrary constants by the inverse $\zeta(\gamma)$ of the shear-viscosity function $\tau(\kappa)$. Conversely, if $\omega(r)$ and $u(r)$ satisfy (9.30), the helical flow may be produced by the aid of suitable boundary tractions in the fluid whose shear-viscosity function is $\tau(\kappa)$.

<u>Exercise 9.4</u>. Prove from (9.25) and (9.27) that

$$T\langle rr\rangle - T\langle zz\rangle = \hat{\sigma}_1(\gamma) - \beta^2\hat{\sigma}_2(\gamma) ,$$

$$T\langle\theta\theta\rangle - T\langle zz\rangle = (\alpha^2 - \beta^2)\hat{\sigma}_2(\gamma) , \tag{9.31}$$

$$T\langle rr\rangle = \rho\upsilon + \int \left\{\frac{1}{r}\left[\alpha^2\hat{\sigma}_2(\gamma) - \hat{\sigma}_1(\gamma)\right] - \rho r\omega(r)^2\right\}dr$$

$$+ az + d\theta + g(t) ,$$

where

$$\hat{\sigma}_r(\gamma) \equiv \sigma_r[\zeta(\gamma)] , \quad r = 1, 2. \tag{9.32}$$

We shall now interpret these results in two major special cases.

Flow Between Rotating Cylinders

In $(9.30)_2$, set $a = b = 0$; then $\beta = 0$, and we may take $u(r) \equiv 0$ in (9.16). The fluid particles move in concentric circles with angular speeds $\omega(r)$ given by $(9.30)_1$. In order that the radial stress $T\langle rr\rangle$ be single-valued, it is necessary by $(9.31)_3$ that $d = 0$. By (9.28),

$$\gamma(r) = \frac{c}{r^2} . \tag{9.33}$$

The one remaining arbitrary constant c is easily interpreted, since the torque M per unit height applied to the cylinder $r =$ const. is given by $M = (2\pi r) r\, T\langle r\theta\rangle$, which by $(9.26)_1$ is $2\pi c$. That is, $c = M/(2\pi)$. This torque M is to be so adjusted that the cylinders $r = R_1$ and $r = R_2$ move with prescribed angular speeds Ω_1 and Ω_2:

$$\omega(R_1) = \Omega_1 , \quad \omega(R_2) = \Omega_2 . \tag{9.34}$$

By $(9.30)_1$,

$$\Omega_2 - \Omega_1 = \int_{R_1}^{R_2} \frac{1}{r} \zeta\left(\frac{M}{2\pi r^2}\right) dr . \tag{9.35}$$

A flow of this kind is often called "Couette flow". Such a flow is approximated in a common type of viscometer, in which the torque applied to one cylinder is measured as a function of the difference of angular velocities. For any simple fluid, the corresponding relation is given by (9.35). If (9.35) is inverted, the inverse ζ of the shear-viscosity function τ is determined as a function of M for given $\Omega_2 - \Omega_1$.

While according to the classical theory the surfaces $z =$ const. sustain an almost uniform pressure when $\upsilon = 0$ and $\rho\omega^2$ is negligibly small, we see from (9.31) that in a general fluid the normal traction $T\langle zz\rangle$ is a function of r. If this traction is not supplied, as for example on the top surface of the fluid in a Couette viscometer, the free surface will tend to rise or fall, according to the signs of the normal-stress functions.

Exercise 9.5. Let the flow be thought of as terminated by a free surface $z = \text{const.}$ on which the atmospheric pressure is a constant, p_0. Balance of total force requires that

$$2\pi \int_{R_1}^{R_2} T\langle zz\rangle \, r dr = -p_0 \pi (R_2^2 - R_1^2) . \tag{9.36}$$

Use this relation to evaluate $g(t)$ in $(9.31)_3$. Denoting the excess of p_0 over the normal pressure $-T\langle zz\rangle$ by N, show that

$$\partial_r N = -\rho r \omega(r)^2 + \frac{1}{r} [\hat{\sigma}_2(\frac{M}{2\pi r^2}) - \hat{\sigma}_1(\frac{M}{2\pi r^2})]$$

$$+ \frac{M}{\pi r^3} \hat{\sigma}_1'(\frac{M}{2\pi r^2}) . \tag{9.37}$$

Flow in a Circular Pipe.

For a second special case of helical flow, we now consider a flow straight down a cylindrical pipe of infinite length, and we assume the fluid adheres to the wall, $r = R$. In (9.30) we take $c = d = 0$, so that $\alpha = 0$, and to keep the velocity finite at $r = 0$ we take also $b = 0$. Then by (9.28)

$$\gamma(r) = \frac{1}{2} ra , \tag{9.38}$$

and a is the specific driving force. From (9.30) we obtain the velocity profile:

$$u(r) = \int_r^R \zeta(\frac{1}{2} a\xi) d\xi , \tag{9.39}$$

whence it is plain that the classical parabolic form holds if and only if ζ is linear. The discharge D is given by

$$\begin{aligned} D(a,R) &= 2\pi \int_0^R r dr \int_r^R \zeta(\frac{1}{2} a\xi) d\xi , \\ &= \pi \int_0^R r^2 \zeta(\frac{1}{2} ar) dr . \end{aligned} \tag{9.40}$$

Thus, the famous "Hagen-Poiseuille formula" or "law of the fourth power" is valid if and only if the shear-viscosity function $\tau(\kappa)$ is linear. If it is found to be valid, we have no assurance that even the Navier-Stokes theory of viscometry is justified, since the nature of the normal-stress functions has no effect on the discharge and hence cannot be determined from measurements of it.

The presence of a radial tension $T\langle rr\rangle$ which is generally different from the other normal tensions suggests that a column of fluid emerging after flowing through a long pipe will tend to swell or shrink in diameter.

Exercise 9.6. Evaluate $g(t)$ in (9.31) by supposing that the total axial force at the exit cross-section be that exerted by a uniform atmospheric pressure p_0. Let P be the excess of that pressure over the radial pressure $-T\langle rr\rangle$ at the exit cross section. Prove that

$$P = az + \frac{1}{R^2}\int_0^R r[\hat{\sigma}_1(\tfrac{1}{2}ar) - \hat{\sigma}_2(\tfrac{1}{2}ar)]dr \quad . \tag{9.41}$$

Hence a sufficient condition that the fluid shall swell upon emergence is

$$2\hat{\sigma}_2(\gamma) - \hat{\sigma}_1(\gamma) > 0 \quad . \tag{9.42}$$

Other Viscometric Flows.

There are other interesting special cases of helical flow. Also two other kinds of viscometric flow have been studied: torsional flow, given in cylindrical co-ordinates by the velocity field

$$\dot{r} = 0 \ , \quad \dot{\theta} = \omega(z) \ , \quad \dot{z} = 0 \ , \tag{9.43}$$

and cone-and-plate flow, given in spherical polar co-ordinates by the velocity field

$$\dot{r} = 0 \quad , \quad \dot{\theta} = 0 \ , \quad \dot{\phi} = \omega(\theta) \quad . \tag{9.44}$$

Since these flows are viscometric, the stress system is easily expressed in terms of the viscometric funcitions. Neither, however, can satisfy the dynamical equations exactly unless non-conservative body forces be supplied. To make them agree roughly with Cauchy's first law when $\underset{\sim}{b} = \underset{\sim}{0}$, it is necessary to suppose the accelerations negligible, and for cone-and-plate flow it is further necessary to suppose θ limited to a very small interval about $\theta = 0$.

In any viscometric flow, the tensor $\underset{\sim}{S}$ is completely determined by κ, $\underset{\sim}{N}$, and the three viscometric functions τ σ_1, and σ_2. Thus all phenomena in the entire class of viscometric flows are simply related to one another. The only problem comes in adjusting κ and $\underset{\sim}{N}$ in such a way as to make the flow dynamically possible, subject to boundary tractions and conservative body force. In this lecture we have considered in detail the two special classes of viscometric flows for which this adjustment is known to be possible exactly, and I have mentioned the other two in which it is known that an approximate solution is possible. In each case, $\underset{\sim}{N}$ and κ were shown to depend upon the nature of the function $\tau(\kappa)$. To find all viscometric flows that are compatible for all simple fluids is an unsolved problem.

LECTURE 10. IMPOSSIBILITY OF RECTILINEAR FLOW IN PIPES.

Problem.

In Lecture 8 the most general form of the stress possible in a simple fluid undergoing viscometric flow was determined, and in Lecture 9 certain classes of viscometric flows were shown to be dynamically possible without bringing to bear non-conservative body force. In these classes, the streamlines are the same as for a Navier-Stokes fluid in the same circumstances, but the distribution of speeds upon them is different, being determined by the shear-viscosity function $\tau(\kappa)$ of the fluid.

Up to the present, very few fixed or simply moving boundaries are known to correspond to flows for which the Navier-Stokes equations can be solved exactly. Nearly all of these give rise to viscometric flows. Those for which the analysis is elementary are exhausted by the cases analysed, at least in outline, in the last lecture. The next easiest class is defined by flow in an infinitely long tube of cross section A, a simply-connected region of the plane. The customary procedure begins by assuming the motion to be an accelerationless lineal flow in the direction of a unit vector $\underset{\sim}{k}$ normal to the cross-section A:

$$\dot{\underset{\sim}{x}} = v(\underset{\sim}{p})\underset{\sim}{k}, \quad \text{where } v(\underset{\sim}{p}) = \underset{\sim}{0} \quad \text{if } \underset{\sim}{p} \in \partial A \ , \tag{10.1}$$

$\underset{\sim}{p}$ being the position vector of a point in the plane. The particles are thus assumed to move at constant speed down the lines parallel to the generators of the pipe wall. The assumption (10.1) is easily shown to be compatible with the Navier-Stokes equations, a partial differential equation for the function $v(\underset{\sim}{p})$ is derived, and it is proved to have a unique solution corresponding to the boundary condition $(10.1)_2$. Thus a unique rectilinear solution exists. Whether it is the only solution of the problem is a far more difficult matter, today unsettled.

We shall now approach the problem in the same spirit for general incompressible fluids. We shall show that in general, no rectilinear flow exists. For proof it suffices to refer to a particular case treated by ERICKSEN, who discovered this remarkable fact. A single counter-example, of course, disproves a general assertion. However, use of the apparatus set up in the previous two lectures makes it easier to see just why no rectilinear flow can be expected in general and to characterize the special cases when rectilinear flow is possible.

Explicit Constitutive Equation.

In the cross-section $\mathcal{A}$, let the curves θ = const. be the orthogonal trajectories of the isovels, v = const., so that a three-dimensional co-ordinate system is given by

$$x_1 = v(\underset{\sim}{p}), \quad x_2 = z, \quad x_3 = \theta(\underset{\sim}{p}) \tag{10.2}$$

where z is the distance from some fixed cross-section, and

$$\nabla\theta \cdot \nabla v = 0. \tag{10.3}$$

(Since only functions of place $\underset{\sim}{x}$ are used here, the longer symbol "grad" for the gradient operator need not be used.) The covariant and contravariant components g_{km} and g^{km} of the unit tensor are

$$g^{11} = g_{11}^{-1} = (\nabla v)^2, \quad g^{22} = g_{22} = 1, \quad g^{33} = g_{33}^{-1} = (\nabla\theta)^2. \tag{10.4}$$

Exercise 10.1. Prove that the flow is viscometric, that

$$\kappa = |\nabla v|, \tag{10.5}$$

and that $\underset{\sim}{N}$ has the form (8.7) with respect to an orthonormal basis tangent to the co-ordinate curves.

Because of this result, we may apply (8.14) and obtain

$$T\langle vz\rangle = \tau(\kappa), \quad T\langle z\theta\rangle = 0, \quad T\langle v\theta\rangle = 0,$$

$$T\langle vv\rangle - T\langle \theta\theta\rangle = \sigma_1(\kappa), \tag{10.6}$$

$$T\langle zz\rangle - T\langle \theta\theta\rangle = \sigma_2(\kappa).$$

Let $\mathbf{i}$ and $\mathbf{j}$ be unit vectors in the directions of ∇v and $\nabla\theta$, so that in particular

$$\nabla v = \mathbf{i}\kappa, \tag{10.7}$$

and

$$\mathbf{T} = T\langle vz\rangle\,(\mathbf{i}\otimes\mathbf{k} + \mathbf{k}\otimes\mathbf{i}) + T\langle vv\rangle\,(\mathbf{i}\otimes\mathbf{i})$$

$$+\, T\langle zz\rangle\,(\mathbf{k}\otimes\mathbf{k}) + T\langle\theta\theta\rangle\,(\mathbf{j}\otimes\mathbf{j}). \tag{10.8}$$

Replacing $\mathbf{j}\otimes\mathbf{j}$ in the last term by $\mathbf{1} - \mathbf{i}\otimes\mathbf{i} - \mathbf{k}\otimes\mathbf{k}$ and using (10.6), we find that

$$\mathbf{T} = (\tau\mathbf{i} + \sigma_2\mathbf{k})\otimes\mathbf{k} + \mathbf{k}\otimes\tau\mathbf{i} + \kappa\mathbf{i}\otimes\frac{\sigma_1}{\kappa}\,\mathbf{i} + T\langle\theta\theta\rangle\,\mathbf{1}, \tag{10.9}$$

where the argument κ of τ, σ_1, and σ_2 is not written. The last term is not determined by the flow.

Dynamical Equation.

Exercise 10.2. Prove that

$$\kappa\nabla\kappa = \kappa\nabla(\kappa\mathbf{i})\mathbf{i}. \tag{10.10}$$

Using this fact and the identity

$$\operatorname{div}(\mathbf{u}\otimes\mathbf{v}) = (\nabla\mathbf{u})\mathbf{v} + \mathbf{u}\operatorname{div}\mathbf{v}, \tag{10.11}$$

show that

$$\operatorname{div}\mathbf{T} = \mathbf{k}\operatorname{div}(\tilde{\mu}\nabla v) + \nabla v\operatorname{div}\left(\frac{\sigma_1}{\kappa^2}\nabla v\right) + \frac{\sigma_1}{\kappa}\nabla\kappa + \nabla T\langle\theta\theta\rangle, \tag{10.12}$$

where

$$\tilde{\mu}(\kappa) \equiv \frac{\tau(\kappa)}{\kappa} . \tag{10.13}$$

If we set

$$h = T\langle\theta\theta\rangle + \int \frac{\sigma_1(\kappa)}{\kappa} d\kappa - \rho\upsilon, \tag{10.14}$$

then

$$\nabla h = \nabla T\langle\theta\theta\rangle + \frac{\sigma_1(\kappa)}{\kappa} \nabla\kappa + \rho \underset{\sim}{b}, \tag{10.15}$$

where $b = -\text{grad}\ \upsilon$. Accordingly, Cauchy's first law (3.16) assumes the form

$$\underset{\sim}{k}\ \text{div}\ (\tilde{\mu}\nabla v) + \nabla v\ \text{div}\left(\frac{\sigma_1}{\kappa^2} \nabla v\right) + \nabla h = 0. \tag{10.16}$$

From this result, then, ∇h is independent of z. Therefore h is of the form

$$h = za + g(\underset{\sim}{p}), \tag{10.17}$$

where a = const., and (10.16) splits into the following two equations:

$$\text{div}\left(\tilde{\mu}(\kappa)\nabla v\right) = -a, \tag{10.18}$$

and

$$\nabla v\ \text{div}\left(\frac{\sigma_1(\kappa)}{\kappa^2} \nabla v\right) + \nabla g = 0. \tag{10.19}$$

The second equation states that g is constant along the isovels v = const. That is, $g(\underset{\sim}{p}) = f(v(\underset{\sim}{p}))$, and (10.19) becomes

$$\text{div}\left(\frac{\sigma_1(\kappa)}{\kappa^2} \nabla v\right) = -f'(v). \tag{10.20}$$

<u>Compatibility</u>.

For any given fluid, the two functions $\tilde{\mu}(\kappa)$ and $\sigma_1(\kappa)$ are given. Accordingly, we have derived <u>two</u> non-linear partial differential equations,

(10.18) and (10.20), to be satisfied by the two functions $v(\underset{\sim}{p})$ and $f(v)$. The function $v(\underset{\sim}{p})$ is to be found; we may choose $f(v)$ at will, if we can.

In some particular cases, these two equations are compatible. For example, if the fluid is such that

$$\sigma_1(\kappa) = c\kappa^2\tilde{\mu}(\kappa), \tag{10.21}$$

then by the choice $f'(v) = ca$, (10.20) becomes identical with (10.18). This case includes the Navier-Stokes theory, for which $c = 0$, and, more generally, any theory for which $\sigma_1(\kappa) = 0$. In the Navier-Stokes theory, (10.18) becomes $\mu\Delta v = -a$, where μ is the viscosity and a is the specific driving force, both assigned. The equation has a unique solution satisfying the boundary condition $(10.1)_2$.

Exercise 10.3. Prove that if (10.18) and (10.20) are compatible, then

$$T\langle zz\rangle - \rho\upsilon = za + \sigma_2(\kappa) - \int \frac{\sigma_1(\kappa)}{\kappa}\, d\kappa + f(v). \tag{10.22}$$

Hence

$$\partial_z[T\langle zz\rangle - \rho\upsilon] = a. \tag{10.23}$$

Thus a is the specific driving force.

More generally, we expect, though it has not been proved, that (10.18) by itself, with assigned $\tilde{\mu}(\kappa)$, is again sufficient to determine a unique $v(\underset{\sim}{p})$ satisfying the boundary condition. If this is so, then such a v will generally fail to satisfy (10.20). Again there are exceptions. If the curves $v = \text{const.}$ are concentric circles or parallel straight lines, then $\kappa = |\nabla v| = g(v)$, and (10.20) is always satisfied. ERICKSEN has shown that if $\tilde{\mu}$ is analytic and if (10.21) does not hold, there are always solutions of (10.18) for which (10.20) fails to hold, and he has conjectured that in fact (10.18 and (10.20) are compatible if and only if either (10.21) holds or $|\nabla v| = f(v)$.

In summary, for a general fluid rectilinear flow in a tube is possible only for exceptional cross sections; for general cross-sections, only for exceptional fluids.

ERICKSEN conjectured that in the general case, a non-rectilinear solution will exist. A departure from a classical streamline pattern is generally described as a "secondary flow". The secondary flow in this case will be a component of velocity normal to the generators of the pipe, as a result of which the fluid moves along spiraliform streamlines.

We may see in advance that calculation of such a flow will be intricate. Indeed, if we assume that the shear-viscosity function $\tau(\kappa)$ and the normal-stress function $\sigma_1(\kappa)$ may be expanded in series in κ, *e.g.* (8.25), then (10.21) is always satisfied to the second order. The effect of incompatibility, then, must be of at least third order in some parameter whose smallness keeps the shearings small. It is natural to seek such a parameter in the specific driving force $\partial_z[T\langle zz\rangle - \rho\upsilon]$. Unfortunately no general method of solving the problem is now known. A procedure of approximation will be presented in Lecture 25.

PART III. ELASTIC MATERIALS

LECTURE 11: ELASTIC MATERIALS.

Statics of Simple Materials.

If a material is at rest and has been so for all time, the deformation history is a constant for each particle:

$$\underset{\sim}{F}^t(\underset{\sim}{X};\, s) = \underset{\sim}{F}(\underset{\sim}{X}) \quad . \tag{11.1}$$

In this class of histories, therefore, anything determined by $\underset{\sim}{F}^t$ is determined by $\underset{\sim}{F}$. In particular, the constitutive equation (3.26) of a simple material reduces to

$$\underset{\sim}{T} = \underset{\sim}{\mathcal{G}}_{\underset{\sim}{\kappa}}(\underset{\sim}{F},\underset{\sim}{X}) \ , \tag{11.2}$$

where $\underset{\sim}{\mathcal{G}}_{\underset{\sim}{\kappa}}$ is a tensor-valued function, the <u>response function</u> of the material with respect to $\underset{\sim}{\kappa}$. A material whose constitutive equation is of the form (11.2) for <u>all</u> deformation histories, not merely rest histories, is said to be <u>elastic</u>. In such a material, the stress at each particle is uniquely determined by the present deformation from a fixed reference configuration, any one.

What has been shown is, <u>the class of simple materials is indistinguishable from the class of elastic materials by static experiments</u>. In other words, the theory of static elasticity is at the same time the <u>statics of simple materials</u>.

In many books "D'Alembert's Principle" is stated as a rule for obtaining dynamical equations from statical ones by adding "inertial forces." In the present case, the static equations of any simple material take the form $\operatorname{div} \underset{\sim}{\mathcal{G}}(\underset{\sim}{F}) + \rho\underset{\sim}{b} = \underset{\sim}{0}$. Adding "inertial forces" yields $\operatorname{div} \underset{\sim}{\mathcal{G}}(\underset{\sim}{F}) + \rho\underset{\sim}{b} = \rho\ddot{\underset{\sim}{x}}$. These are the correct dynamical equations for elastic materials only. If this popular "D'Alembert's Principle" were correct, all simple materials would have to be elastic.

Exercise 11.1. Prove that

$$\mathfrak{g}_{\kappa_2}(\mathbf{F}) = \mathfrak{g}_{\kappa_1}(\mathbf{F}\mathbf{P}) \quad , \tag{11.3}$$

where $\mathbf{P}$ is the gradient of the deformation that carries κ_1 into κ_2. Hence the definition of an elastic material, which seems to employ a particular κ, is invariant under change of reference configuration.

A similar definition holds for incompressible elastic materials:

$$\mathbf{T} = -p\mathbf{1} + \mathfrak{g}(\mathbf{F}), \qquad |\det \mathbf{F}| = 1 \tag{11.4}$$

where, as usual henceforth, the argument $\mathbf{X}$ and the subscript κ are not written.

Reduction for Frame-indifference.

The functional equation (4.1) expressing frame-indifference reduces for elastic materials to

$$\mathfrak{g}(\mathbf{Q}\mathbf{F}) = \mathbf{Q}\mathfrak{g}(\mathbf{F})\mathbf{Q}^T \ , \tag{11.5}$$

which holds identically in the non-singular tensor $\mathbf{F}$ and the orthogonal tensor $\mathbf{Q}$. The solution of this equation may be read off from (4.5):

$$\mathbf{T} = \mathbf{R}\mathfrak{g}(\mathbf{U})\mathbf{R}^T \ . \tag{11.6}$$

Many other reduced forms are valid.

Elastic Fluids.

An elastic material may be a solid or a fluid crystal. If it is isotropic but not solid, of course it is a fluid. From (6.24) we see at once that the constitutive equation of an elastic fluid is

$$\mathbf{T} = -p(\rho)\mathbf{1} \ , \tag{11.7}$$

while for an incompressible elastic fluid

$$\mathbf{T} = -p\mathbf{1} \ , \tag{11.8}$$

where p is arbitrary. Elastic fluids are studied intensively in hydrodynamics, aerodynamics, meteorology, and oceanography.

The Piola-Kirchhoff Stress Tensor.

For problems concerning elastic non-fluids a description from start to finish in terms of a reference configuration is often useful. To this end we write the resultant contact force on $\mathcal{P}_\chi$ as the integral of a traction vector $\mathbf{t}_\kappa$ over the boundary $\partial\mathcal{P}_\kappa$ of $\mathcal{P}$ in the reference configuration κ:

$$\mathbf{f}_c(\mathcal{P}) = \int_{\partial\mathcal{P}_\chi} \mathbf{t}\, ds(\mathbf{x}) = \int_{\partial\mathcal{P}_\kappa} \mathbf{t}_\kappa\, ds(\mathbf{X}) \quad , \tag{11.9}$$

where $\mathbf{x} = \chi(\mathbf{X})$ as usual. Thus $\mathbf{t}_\kappa$ is parallel to $\mathbf{t}$, but its magnitude is adjusted according to the local change of area:

$$\mathbf{t}_\kappa = \frac{ds(\mathbf{x})}{ds(\mathbf{X})}\,\mathbf{t} \quad . \tag{11.10}$$

Because of Cauchy's fundamental lemma (3.11), there exists a tensor $\mathbf{T}_\kappa(\mathbf{X},t)$ such that

$$\mathbf{t}_\kappa = \mathbf{T}_\kappa \mathbf{n}_\kappa = \frac{ds(\mathbf{x})}{ds(\mathbf{X})}\,\mathbf{T}\mathbf{n} \ , \tag{11.11}$$

where $\mathbf{n}_\kappa$ is the outer unit normal at $\mathbf{X}$ in κ. But from integral calculus

$$\mathbf{n} = J(\mathbf{F}^{-1})^T \mathbf{n}_\kappa \frac{ds(\mathbf{X})}{ds(\mathbf{x})} \quad . \tag{11.12}$$

Hence

$$\mathbf{T}_\kappa = J\mathbf{T}(\mathbf{F}^{-1})^T \ , \quad \mathbf{T} = J^{-1}\mathbf{T}_\kappa \mathbf{F}^T \ . \tag{11.13}$$

The tensor $\mathbf{T}_\kappa$ is the Piola-Kirchhoff stress tensor.

It is important to notice that while $\mathbf{T}$ and $\mathbf{T}_\kappa$ serve to determine the traction upon every part of the body, neither determines the other unless the deformation $\mathbf{F}$ from κ to the present configuration χ be specified. The CAUCHY stress $\mathbf{T}$ delivers the present tractions alone and completely; the PIOLA-KIRCHHOFF stress $\mathbf{T}_\kappa$ delivers those same tractions only if $\mathbf{F}$ is specified.

Exercise 11.2. Prove that Cauchy's laws (3.16) and (3.18) assume the following forms in terms of $\underset{\sim}{T}_{\kappa}$:

$$\text{Div}\, \underset{\sim}{T}_{\kappa} + \rho_{\kappa} \underset{\sim}{b} = \rho_{\kappa} \ddot{\underset{\sim}{x}} \quad , \quad \underset{\sim}{T}_{\kappa} \underset{\sim}{F}^{T} = (\underset{\sim}{T}_{\kappa} \underset{\sim}{F}^{T})^{T} \quad , \tag{11.14}$$

where the capital D on "Div" serves to remind us that $\underset{\sim}{T}_{\kappa}$ is regarded as a function of $\underset{\sim}{X}$ here. The differential equation $(11.14)_1$ is to be satisfied in the interior of the reference configuration $\mathcal{B}_{\kappa}$; the constitutive equation of the material is assumed to be such that $(11.14)_2$ is satisfied identically.

In terms of the Piola-Kirchhoff tensor, the constitutive equation (11.2) of elasticity assumes the form

$$\underset{\sim}{T}_{\kappa} = \underset{\sim}{\mathfrak{h}}_{\kappa}(\underset{\sim}{F}, \underset{\sim}{X}) = \underset{\sim}{\mathfrak{h}}(\underset{\sim}{F}) \quad , \tag{11.15}$$

say, while the functional equation (11.5) expressing frame-indifference becomes

$$\underset{\sim}{\mathfrak{h}}(\underset{\sim}{Q}\underset{\sim}{F}) = \underset{\sim}{Q}\, \underset{\sim}{\mathfrak{h}}(\underset{\sim}{F}) \quad . \tag{11.16}$$

Natural States.

A stress-free configuration is called a natural state. If such a configuration is used as reference,

$$\underset{\sim}{\mathfrak{g}}(\underset{\sim}{1}) = \underset{\sim}{0} \quad , \quad \underset{\sim}{\mathfrak{h}}(\underset{\sim}{1}) = \underset{\sim}{0} \quad . \tag{11.17}$$

In the applications to solids it is customary, though often not necessary, to assume that the elastic material does have a natural state. In fluid mechanics it is customary to impose conditions such as to insure that $p(\rho) > 0$ if $\rho > 0$. If $\rho_{\kappa} \neq 0$ in some κ, $\rho \neq 0$ in all configurations obtainable by smooth deformations, by (1.18). Hence such fluids have no natural states. In these lectures we shall not assume that a material has a natural state without saying so expressly.

Differential Equations and Boundary Conditions.

Substitution of the constitutive equation (11.15) into (11.14) yields a differential equation to be satisfied by the motion $\underset{\sim}{\chi}_{\kappa}$ of the elastic material at all interior points $\underset{\sim}{X}$ in κ:

$$\mathrm{Div}\, \boldsymbol{\mathfrak{h}}(\mathbf{F},\mathbf{X}) + \rho_{\kappa}\mathbf{b} = \rho_{\kappa}\ddot{\mathbf{x}} \, . \tag{11.18}$$

Explicitly, let a co-ordinate form of (11.15) be

$$T_{\kappa m}{}^{\alpha} = \mathfrak{h}_m{}^{\alpha}(x^p{}_{,\beta},\, X^{\gamma}) \, , \tag{11.19}$$

and set

$$A_m{}^{\alpha}{}_p{}^{\beta} \equiv \partial_{x^p{}_{,\beta}} \mathfrak{h}_m{}^{\alpha} \quad \text{or} \quad \mathbf{A}(\mathbf{F}) = \partial_{\mathbf{F}} \boldsymbol{\mathfrak{h}}(\mathbf{F}),$$

$$q_m \equiv \partial_{X^\alpha} \mathfrak{h}_m{}^{\alpha} \, . \tag{11.20}$$

The fourth-order tensor $\mathbf{A}$ is called an <u>elasticity</u> of the elastic material. In terms of it, (11.18) has the co-ordinate form

$$A_m{}^{\alpha}{}_p{}^{\beta}\, x^p{}_{,\alpha;\beta} + q_m + \rho_{\kappa} b_m = \rho_{\kappa}\ddot{x}_m \tag{11.21}$$

Here $x^p{}_{,\alpha;\beta}$ denotes a double covariant derivative. If rectangular Cartesian co-ordinates are used,

$$x^p{}_{,\alpha;\beta} = \partial_{X^\beta}\, \partial_{X^\alpha}\, x^p(\mathbf{X},t) \, . \tag{11.22}$$

Specification of the place $\mathbf{x}$ into which a boundary point $\mathbf{X}$ in κ is deformed constitutes a <u>boundary condition of place</u>:

$$\boldsymbol{\chi}(\mathbf{X},t) = \text{a given function of } \mathbf{X} \text{ and } t \text{ if } \mathbf{X} \in \partial\mathcal{B}_{\kappa} \, . \tag{1.23}$$

An appropriate boundary condition of traction is harder to formulate. We wish to find the deformation produced by applying given forces to a given body in κ. We cannot simply specify the tractions acting upon $\partial\mathcal{B}_{\kappa}$ in the present configuration $\boldsymbol{\chi}$, since that configuration itself is unknown. If we merely specify the traction on the boundary in κ, the deformation produced will move the point of application and deform the surface, so the prescription may not correspond to any natural concept of applying given forces. The boundary condition of traction customarily imposed is

$$\underset{\sim}{t}_{\underset{\sim}{\kappa}} = \text{a given function of } \underset{\sim}{X} \text{ and } t \text{ if } \underset{\sim}{X} \in \partial \mathcal{B}_{\underset{\sim}{\kappa}} \cdot \qquad (11.24)$$

This condition may be described as expressing a strategy for stating an unequivocal boundary-value problem. By (11.10), the direction of $\underset{\sim}{t}$ is thus prescribed, but not its magnitude. To visualize the application of this condition, suppose that assigned tractions act on $\partial \mathcal{B}_{\underset{\sim}{\kappa}}$, and in the resulting deformation keep the directions of those tractions fixed but adjust their magnitudes in the ratio $ds(\underset{\sim}{X})/ds(\underset{\sim}{x})$. It is the force per unit initial area that is prescribed.

Non-Uniqueness of Solution.

No general uniqueness theorem is known at present in elasticity theory. Certainly, no unqualified assertion of uniqueness can be true.

Consider, for example, an elastic material confined between concentric cylinders. Keeping the inner cylindrical boundary fixed, rotate the outer one counter-clockwise through a straight angle, so that each point on it goes into its antipodal point. The material within is expected to undergo some smooth deformation. If, instead, the same rotation of the boundary had been effected clockwise, the points on the outer boundary would experience the same displacement, but the interior points, being pulled clockwise instead of counter-clockwise, would certainly have experienced a different one. Thus no reasonable theory of elasticity can be expected to give a unique solution to this boundary-value problem of place. In fact, a little thought shows that there ought to be infinitely many distinct solutions corresponding to any rotation, large or small.

Turning now the boundary-value problem of traction, consider a hemispherical bowl subject to no traction at all. Experience teaches us that such a bowl, if not too thick, may be snapped inside out and will remain in its everted shape subject to no traction at all. No good theory of finite elasticity should exclude this eversion. Indeed, it is a central problem to calculate the new shape according to the theory. Thus unqualified uniqueness to the boundary-value of traction is not desirable.

This point may be strengthened through a different example. We do expect uniqueness in one case, namely, homogeneous extension or contraction of a bar subject to a uniaxial tension t or a pressure $-t$. However, the boundary-value problem of traction specifies, not this traction t, but the traction $\underset{\sim}{t}_{\underset{\sim}{\kappa}}$ per unit area in the reference configuration $\underset{\sim}{\kappa}$. If tensions $t_{\underset{\sim}{\kappa}}$ act outward upon the plane ends of a bar in $\underset{\sim}{\kappa}$, there are two deformed configurations corresponding to this prescription: In one, the $t_{\underset{\sim}{\kappa}}$ still act outward, as tensions, upon the deformed ends; in the other, they have undergone a rotation through a straight angle, and since their directions are fixed, they have become compressions. The actual tractions t are different in the two cases, since in one case the area of the ends will decrease, while in the other it will increase. We expect both these problems to have a solution, and we expect no others. That is, this particular boundary-value problem of place, like that of eversion, should have exactly two solutions.

Indeed a,uniqueness theorem of some sort must hold, but as yet no adequate formulation of such a theorem, let alone a proof of it, has been given.

Existence and Uniqueness in the Limit of Vanishing Loads.

A theorem of existence and uniqueness has been proved by STOPPELLI for a particular class of boundary-value problems of traction. Letting $\underset{\sim}{b}$ and $\underset{\sim}{t}_{\underset{\sim}{\kappa}}$ be fixed function of position in $\mathcal{P}_{\underset{\sim}{\kappa}}$ and $\partial\mathcal{P}_{\underset{\sim}{\kappa}}$, respectively, he considers body force $\varepsilon\underset{\sim}{b}$ and boundary traction $\varepsilon\underset{\sim}{t}$. He proves that for sufficiently small ε, a unique solution exists, provided the loads do not possess an axis of equilibrium. (If a system of forces in equilibrium can be rotated through any angle about an axis and yet maintain equilibrium of moments, that axis is called an axis of equilibrium. For such a load system, clearly, a rotation will not change the state of stress, so a unique deformation cannot be expected.) STOPPELLI proves further that the solution is an analytic function of ε and hence may be calculated by a method of power series.

STOPPELLI's theorems give a position to the classical infinitesimal theory of elasticity as an approximation to the general theory, and a means of improving that approximation, but they do not illuminate problems of large deformation.

Isotropic Elastic Materials.

By (6.13) an isotropic elastic material has a constitutive equation of the form

$$\mathbf{T} = \mathfrak{f}(\mathbf{B}) , \tag{11.25}$$

where by (6.14)

$$\mathfrak{f}(\mathbf{QBQ})^T = \mathbf{Q}\,\mathfrak{f}(\mathbf{B})\mathbf{Q}^T , \tag{11.26}$$

for all orthogonal $\mathbf{Q}$ and all symmetric $\mathbf{B}$.

Exercise 11.3. Without recourse to (6.13) and (6.14), prove that (11.2) is equivalent to (11.25) and (11.26) if and only if the elastic material is isotropic and the reference configuration is an undistorted state.

Representation of Isotropic Functions.

Functions satisfying (11.26) are called isotropic. According to a famous theorem due to RIVLIN & ERICKSEN, a function $\mathfrak{f}(\mathbf{A})$ whose values and arguments are symmetric tensors in n dimensions is isotropic if and only if it has a representation of the form

$$\mathbf{D} \equiv \mathfrak{f}(\mathbf{A}) = \phi_0\mathbf{1} + \phi_1\mathbf{A} + \ldots + \phi_{n-1}\mathbf{A}^{n-1} , \tag{11.27}$$

where the ϕ_i are invariant functions of $\mathbf{A}$:

$$\phi_i(\mathbf{QAQ}^T) = \phi_i(\mathbf{A}) . \tag{11.28}$$

The following proof is due to SERRIN. Let $\mathbf{e}$ be a proper vector of $\mathbf{A}$, and let $\mathbf{Q}$ be a reflection in the plane normal to $\mathbf{e}$:

$$\mathbf{Qe} = -\mathbf{e} , \quad \mathbf{Qf} = \mathbf{f} \quad \text{if} \quad \mathbf{f}\cdot\mathbf{e} = 0 . \tag{11.29}$$

Then $\mathbf{QAQ}^T = \mathbf{A}$, and by hypothesis $\mathbf{QDQ}^T = \mathbf{D}$. Therefore

$$\mathbf{Q}(\mathbf{De}) = \mathbf{D}(\mathbf{Qe}) = -\mathbf{De} . \tag{11.30}$$

That is, $\mathbf{Q}$ transforms $\mathbf{De}$ into its opposite. This can be so, by (11.29), if and only if $\mathbf{De}$ is parallel to $\mathbf{e}$:

$$\underset{\sim}{D}\underset{\sim}{e} = d\underset{\sim}{e} . \tag{11.31}$$

Thus $\underset{\sim}{e}$ is also a proper vector of $\underset{\sim}{D}$.

Let the distinct proper numbers of $\underset{\sim}{A}$ be $a_1, a_2, \ldots, a_m$ with proper vectors $\underset{\sim}{e}_1, \underset{\sim}{e}_2, \ldots, \underset{\sim}{e}_m$, $m \leqq n$. The $\underset{\sim}{e}_i$ are proper vectors of $\underset{\sim}{D}$. Let the corresponding proper numbers, which need not be distinct, be d_i. Consider the following linear system for quantities $\phi_0, \phi_1, \ldots, \phi_m$:

$$d_k = \phi_0 + \phi_1 a_k + \phi_2 a_k + \ldots + \phi_{m-1} a_k^{m-1} , \qquad k = 1, 2, \ldots, m. \tag{11.32}$$

The determinant of this system, being $\prod_{j \leqq k}(a_j - a_k)$, is not zero, so unique ϕ_i exist for given d_k and a_k. Hence

$$\underset{\sim}{D} = \phi_1 \underset{\sim}{1} + \phi_1 \underset{\sim}{A} + \ldots + \phi_{m-1} \underset{\sim}{A}^{m-1} . \tag{11.33}$$

Since $$\underset{\sim}{\mathfrak{f}}(\underset{\sim}{A}) = \underset{\sim}{\mathfrak{f}}(\underset{\sim}{Q}\underset{\sim}{A}\underset{\sim}{Q}^T) = \underset{\sim}{1}\phi_0 + (\underset{\sim}{Q}\underset{\sim}{A}\underset{\sim}{Q}^T)\phi_1 + \ldots + (\underset{\sim}{Q}\underset{\sim}{A}\underset{\sim}{Q}^T)^{m-1}\phi_m , \tag{11.34}$$

the ϕ_i for $\underset{\sim}{A}$ are also the ϕ_i for $\underset{\sim}{Q}\underset{\sim}{A}\underset{\sim}{Q}^T$. Thus (11.28) is satisfied.

If the proper numbers of $\underset{\sim}{A}$ are all distinct, $m = n$, and the proof is complete. If not, we can take $\phi_m = \phi_{m+1} = \ldots = \phi_{n-1} = 0$, and again the proof of (11.27) is complete, but this is only one possible choice for the ϕ_i, which are not uniquely determined.

The coefficient functions ϕ_i do not generally inherit whatever smoothness properties $\underset{\sim}{\mathfrak{f}}(\underset{\sim}{A})$ may have. For example, if $\underset{\sim}{\mathfrak{f}}(\underset{\sim}{A})$ is continuous, the ϕ_i need not be.

Exercise 11.4. Prove that a scalar function ϕ_i satisfies (11.28) if and only if

$$\phi_i = \phi_i(I_1, I_2, \ldots, I_n) , \tag{11.35}$$

where $I_1, I_2, \ldots, I_n$ are the principal invariants of $\underset{\sim}{A}$, viz, the elementary symmetric functions of the proper numbers $a_1, a_2, \ldots, a_n$.

Returning to elasticity, we have now a representation theorem for the constitutive equation (11.25), referred to an undistorted state:

$$\underset{\sim}{T} = \aleph_0 \underset{\sim}{1} + \aleph_1 \underset{\sim}{B} + \aleph_2 \underset{\sim}{B}^2 \tag{11.36}$$

In view of the Hamilton-Cayley equation $\underset{\sim}{B}^3 - I\underset{\sim}{B}^2 + II\underset{\sim}{B} - III\ \underset{\sim}{1} = \underset{\sim}{0}$, where I, II, and III are the principal invariants of $\underset{\sim}{B}$, we may express (11.36) in the alternative form

$$\underset{\sim}{T} = \beth_0 \underset{\sim}{1} + \beth_1 \underset{\sim}{B} + \beth_{-1} \underset{\sim}{B}^{-1} \quad . \tag{11.37}$$

The response coefficients $\aleph_0, \ldots, \beth_{-1}$ are functions of I, II, and III.

Included in the result just proved are the following two theorems: <u>Every principal axis of strain in $\underset{\sim}{\chi}$ is also a principal axis of stress</u>, and <u>the stress on an undistorted state of an isotropic elastic material is always hydrostatic</u>.

<u>Principal Stresses and Principal Forces.</u>

The proper numbers t_i of $\underset{\sim}{T}$ are called <u>principal stresses</u>. In an isotropic material, by (11.36) they are related to the principal stretches v_i by

$$\begin{aligned} t_i &= \aleph_0 + \aleph_1 v_i^2 + \aleph_2 v_i^4 , \\ &= t_i\,(v_1, v_2, v_3) , \end{aligned} \tag{11.38}$$

where $i = 1, 2, 3$.

<u>Exercise 11.5</u>. Three functions $t_i(v_1, v_2, v_3)$ of non-negative arguments define an isotropic elastic material by (11.38) if and only if there exists a function $\bar{t}(\xi,\eta,\rho)$ such that

$$\begin{aligned} \bar{t}(\xi,\eta,\zeta) &= \bar{t}(\xi,\zeta,\eta) , \\ t_1 &= \bar{t}(v_1, v_2, v_3) , \\ t_2 &= \bar{t}(v_2, v_3, v_1) , \\ t_3 &= \bar{t}(v_3, v_1, v_2) . \end{aligned} \tag{11.39}$$

If we substitute the polar decomposition $\underset{\sim}{F} = \underset{\sim}{V}\underset{\sim}{R}$ in (11.13), we obtain

$$\underset{\sim}{T}_{\underset{\sim}{\kappa}} = (J\underset{\sim}{T}\underset{\sim}{V}^{-1})\underset{\sim}{R} \quad . \tag{11.40}$$

In an isotropic elastic material, $\underset{\sim}{T}$ and $\underset{\sim}{V}^{-1}$ are co-axial. The proper numbers of the symmetric tensor $J\underset{\sim}{T}\underset{\sim}{V}^{-1}$ are given by

$$T_i = v_1 v_2 v_3 \frac{t_i}{v_i} , \qquad i = 1, 2, 3 \ . \tag{11.41}$$

These are called the <u>principal forces</u> of the stress system with respect to $\underset{\sim}{\kappa}$, since they correspond to the same resultant force on a specific area in the reference configuration as do the principal stresses acting on the corresponding area in the present configuration. Often the T_i are called "engineering stresses".

<u>Differential Equations for Isotropic Materials</u>.

If we substitute (11.25) in Cauchy's first law (3.16), we obtain

$$F^{km}{}_{pq} B^{pq}{}_{,m} + \rho b^k = \rho \ddot{x}^k \tag{11.42}$$

where

$$F^{km}{}_{pq} = \partial_{B^{pq}} \tilde{f}^{km}(\underset{\sim}{B}) \tag{11.43}$$

Here the independent variables are $\underset{\sim}{x}$ and t, so that the entire problem may be seen in terms of the deformed, loaded configuration. The tensor F is also called an <u>elasticity</u> of the material.

Components with respect to an orthonormal triad of proper vectors of $\underset{\sim}{B}$ (which, of course, are also proper vectors of $\underset{\sim}{T}$) are called <u>principal components</u>. <u>E.g.</u>, the v_i and t_i are principal components.

Exercise 11.6. The principal components of F are of the types

$$F\langle 11\ 11\rangle = \partial_{v_1^2} t_1 \,,$$

$$F\langle 11\ 22\rangle = \partial_{v_2^2} t_1 \,, \ \ldots, \tag{11.44}$$

$$F\langle 12\ 12\rangle = \frac{t_1 - t_2}{2(v_1^2 - v_2^2)} \,, \ \ldots \,,$$

while other types of components vanish. Interpret the 9 non-vanishing principal elasticities as tangent moduli of extension and secant moduli of shear.

Incompressible Materials.

For an incompressible isotropic elastic material, by (11.4) and (11.27), counterparts of (11.36) and (11.37) may be written down:

$$\mathbf{T} = -p\mathbf{1} + \aleph_1 \mathbf{B} + \aleph_2 \mathbf{B}^2 \,,$$

$$= -p\mathbf{1} + \beth_1 \mathbf{B} + \beth_{-1} \mathbf{B}^{1} \,, \tag{11.45}$$

where p is arbitrary, where $\aleph_1, \ldots, \beth_{-1}$ are functions of I and II, and where only deformations such that $\det \mathbf{B} = 1$ are allowed.

LECTURE 12. NORMAL TRACTIONS IN SIMPLE SHEAR OF ISOTROPIC BODIES.

Problem.

Homogeneous motions of homogeneous simple bodies have been studied in Lecture 5. From the results there it follows as a special case that any static homogeneous deformation may be maintained by the effect of boundary tractions alone in any homogeneous simple body. Such a deformation may carry the body from any one homogeneous configuration into any other.

The particular homogeneous deformation called simple shear serves as illustration of the type of tractions required. Such a deformation is given by (5.11) at any one fixed instant: $\kappa t = K$, say. Thus

$$x_1 = X_1$$
$$x_2 = X_2 + KX_1, \tag{12.1}$$
$$x_3 = X_3.$$

The constant K is called the amount of shear.

Since

$$[\underset{\sim}{F}] = \begin{Vmatrix} 1 & 0 & 0 \\ K & 1 & 0 \\ 0 & 0 & 1 \end{Vmatrix}, \tag{12.2}$$

it follows that

$$[\mathbf{B}] = [\mathbf{F}\mathbf{F}^T] = \begin{Vmatrix} 1 & K & 0 \\ K & 1+K^2 & 0 \\ 0 & 0 & 1 \end{Vmatrix},$$

$$[\mathbf{B}^{-1}] = \begin{Vmatrix} 1+K^2 & -K & 0 \\ -K & 1 & 0 \\ 0 & 0 & 1 \end{Vmatrix}, \tag{12.3}$$

$$I = \operatorname{tr} \mathbf{B} = 3 + K^2 = II = \operatorname{tr} \mathbf{B}^{-1}, \qquad III = 1.$$

In the theory of infinitesimal elastic deformation from a natural state, a simple shear may be effected by applying a proportional shear stress in the same direction. In an isotropic elastic body in finite strain such a simple correlation of effects is no longer possible. As we shall see, shear stress alone can never produce simple shear.

We shall consider only homogeneous isotropic bodies referred to a homogeneous configuration.

Unconstrained Bodies.

To find the stresses required to effect the simple shear (12.1) in an unconstrained body, we substitute (12.3) into (11.37). If we set

$$\hat{\beth}_r \equiv \hat{\beth}_r(K^2) \equiv \beth_r(3 + K^2,\ 3 + K^2,\ 1), \tag{12.4}$$

the result is

$$[\mathbf{T}] = (\hat{\beth}_0 + \hat{\beth}_1 + \hat{\beth}_{-1})[\mathbf{1}] + K(\hat{\beth}_1 - \hat{\beth}_{-1}) \begin{Vmatrix} 0 & 1 & 0 \\ 1 & 0 & 0 \\ 0 & 0 & 0 \end{Vmatrix}$$
$$+ K^2 \beth_1 \begin{Vmatrix} 0 & 0 & 0 \\ 0 & 1 & 0 \\ 0 & 0 & 0 \end{Vmatrix} + K^2 \beth_{-1} \begin{Vmatrix} 1 & 0 & 0 \\ 0 & 0 & 0 \\ 0 & 0 & 0 \end{Vmatrix}. \tag{12.5}$$

In particular, the relation between shear stress $T\langle 12\rangle$ and amount of shear K is

$$T\langle 12\rangle = K\hat{\mu}(K^2), \tag{12.6}$$

where

$$\hat{\mu}(K^2) \equiv \hat{\beth}_1(K^2) - \hat{\beth}_{-1}(K^2). \tag{12.7}$$

Thus, as expected, $T\langle 12\rangle$ is an odd function of K, and any such function is compatible with the theory based on (11.37). The function $\hat{\mu}(K^2)$ is the generalized <u>shear modulus</u> of the material in the particular undistorted state used as reference. The classical shear modulus $\mu \equiv \hat{\mu}(0)$. If the response coefficients $\beth_1(I,\ II,\ III)$ are differentiable at the argument $(1,\ 1,\ 1)$,

$$\hat{\mu}(K^2) = \mu + O(K^2). \tag{12.8}$$

Therefore any departure from the classical proportionality of shear stress to the amount of shear produced is an effect of at least third order in the latter.

<u>Exercise 12.1</u>. In simple shear

$$v_1^2 = 1 + \frac{1}{2}K^2 + K\sqrt{1 + \frac{1}{4}K^2}\ ,$$

$$v_2^2 = 1 + \frac{1}{2}K^2 - K\sqrt{1 + \frac{1}{4}K^2} = \frac{1}{v_1^2}\ , \tag{12.9}$$

$$v_3^2 = 1,$$

Hence

$$\hat{\mu}(K^2) = \frac{t_1 - t_2}{v_1^2 - v_2^2}\ . \tag{12.10}$$

For an elastic fluid, $\hat{\mu}(K^2) = 0$ for all K. While it is not known whether any non-fluids can satisfy this condition, in the interpretation of results we shall exclude any such material.

Returning to (12.5), we see that in order for shear stress to suffice for maintenance of simple shear, it would be necessary that $\beth_1 - \beth_{-1} = 0$, so $\hat{\mu}(K^2) = 0$ by (12.7). Thus the shear stress also would vanish. That is, the response predicted by the infinitesimal theory cannot hold exactly for <u>any</u> isotropic elastic material with non-vanishing shear modulus. In general, normal tractions on the planes of shear $X_3 =$ const., the shearing planes $X_1 =$ const., and the normal planes $X_2 =$ const. must be supplied in order for the shear to be effected.

From (12.5) these tractions are easily written down:

$$\frac{T\langle 33\rangle}{K^2} \equiv \tau(K^2) = \frac{\hat{\beth}_0 + \hat{\beth}_1 + \hat{\beth}_{-1}}{K^2},$$

$$\frac{T\langle 11\rangle - T\langle 33\rangle}{K^2} = \hat{\beth}_{-1}(K^2), \tag{12.11}$$

$$\frac{T\langle 22\rangle - T\langle 33\rangle}{K^2} = \hat{\beth}_1(K^2).$$

It is impossible that all three normal tractions be equal, or even that $T\langle 11\rangle = T\langle 22\rangle$, for again $\hat{\mu}(K^2) = 0$ for all shears, a case we have agreed to set aside. Thus of the normal tractions required to effect the simple shear, at most two can be equal. Indeed, from (12.11) and (12.6) we find the following <u>universal relation</u>:

$$T\langle 22\rangle - T\langle 11\rangle = KT\langle 12\rangle, \tag{12.12}$$

a relation connecting three of the components of $\mathbf{T}$, <u>no matter what</u> be the response coefficients $\beth_r$. A relation of this kind serves as a check on the theory as a whole. If (12.12) is not satisfied in a particular case, then no choice of the response coefficients $\beth_r$ in (11.37) can force the case in question into conformity with the theory of isotropic elastic materials.

<u>Exercise 12.2</u>. Let N and T be the normal and tangential tractions on the deformed configuration of a plane $X_2 =$ const. Prove that

$$(1 + K^2)T = K(T\langle 22\rangle - T\langle 11\rangle) + (1 - K^2)T\langle 12\rangle,$$

$$(1 + K^2)N = T\langle 22\rangle - 2KT\langle 12\rangle + K^2 T\langle 11\rangle. \tag{12.13}$$

Hence the universal relation (12.12) may be expressed in the form

$$KT\langle 12\rangle = (1 + K^2)(T\langle 11\rangle - N). \tag{12.14}$$

If K is small, $N \approx T\langle 11\rangle$. However, if $N = T\langle 11\rangle$, again we conclude that $T\langle 12\rangle = 0$. Thus, rigorously, it is impossible for the two normal tractions $T\langle 11\rangle$ and N to be equal. More generally, if the response coefficients $\hat{\beth}_r$ are differentiable at the reference configuration,

$$\hat{\beth}_r(K^2) = \hat{\beth}_r(0) + O(K^2), \tag{12.15}$$

and by (12.11) the inequality of the normal tractions is an effect, generally, of <u>second order</u> in the amount of shear. The presence of these inequalities is called the <u>Poynting effect</u>. Using a very special theory of elasticity, POYNTING noticed formulae such as (12.11). He inferred that if shear stress without these normal tractions is applied to a block, the faces will draw together or spread apart by an amount proportional to K^2.

The mean tension is given by

$$\frac{1}{3}\,\mathrm{tr}\,\underset{\sim}{T} = [\tau + \frac{1}{3}(\hat{\beth}_1 + \hat{\beth}_{-1})]K^2. \tag{12.16}$$

The presence of such a tension is called the <u>Kelvin effect</u>. If the undistorted reference configuration is a natural state, by (12.5)

$$\hat{\beth}_0(0) + \hat{\beth}_1(0) + \hat{\beth}_{-1}(0) = 0. \tag{12.17}$$

If also (12.15) holds, then $\frac{1}{3}\,\mathrm{tr}\,T = O(K^2)$. KELVIN obtained results of this kind from a very special theory. He inferred that if, conversely, the hydrostatic tension corresponding to (12.16) is not supplied, a sheared isotropic body will tend to contract or expand, according to the sign of (12.16), in proportion to the square of the amount of shear.

<u>Incompressible Bodies</u>.

Since simple shear is isochoric, it may be effected also in incompressible bodies, although the stress system required will be of a somewhat different kind. By (11.45) and (12.3),

$$[\underset{\sim}{T}] = -p_0[\underset{\sim}{1}] + \hat{\mu}(K^2) \begin{Vmatrix} 0 & 1 & 0 \\ 1 & 0 & 0 \\ 0 & 0 & 0 \end{Vmatrix}$$

$$+ K^2 \beth_1 \begin{Vmatrix} 0 & 0 & 0 \\ 0 & 1 & 0 \\ 0 & 0 & 0 \end{Vmatrix} + K^2 \beth_{-1} \begin{Vmatrix} 1 & 0 & 0 \\ 0 & 0 & 0 \\ 0 & 0 & 0 \end{Vmatrix}, \tag{12.18}$$

where $\hat{\mu}(K^2)$ is again defined by (12.7) and where $\hat{\beth}_r \equiv \hat{\beth}_r(K^2) \equiv \beth_r(3 + K^2, 3 + K^2)$. The dynamical condition $\operatorname{div} \underset{\sim}{T} = \underset{\sim}{0}$ requires that $p_0 = \text{const}$.

Since the material is incompressible, the Kelvin effect cannot occur. Because of the arbitrary pressure p_0, the variety of Poynting effects possible becomes greater. The pressure p_0 may be chosen so as to leave any one of the families of planes $X_1 = \text{const.}$, $X_2 = \text{const.}$, $x_2 = \text{const.}$, $X_3 = \text{const.}$ free of normal traction. E.g., if we choose $p_0 = 0$, then $T\langle 33\rangle = 0$. In this case

$$\frac{T\langle 11\rangle}{K^2} = \hat{\beth}_{-1}, \quad \frac{T\langle 22\rangle}{K^2} = \hat{\beth}_1, \tag{12.19}$$

affording an immediate interpretation of the response coefficients. Thus, again, $T\langle 11\rangle \neq T\langle 22\rangle$ and, since (12.14) holds, $T\langle 11\rangle \neq N$.

Remarks.

While simple shear is the most illuminating homogeneous static deformation, other cases are important as well, notably simple extension and uniform dilatation. More important than any one case is the fact that homogeneous static deformations are possible in all homogeneous elastic bodies, subject to surface tractions alone. Deformation fields which can be assigned in advance, without solving any differential equations, allow an experimental program for determining the nature of the stress relation. Such a program would be difficult if confined to deformation fields not known in advance to be possible, for the experimenter would have first to conjecture a form for the function he is measuring and solve the resulting problem to find the tractions such as to produce it, then apply those tractions, and check the result. With homogeneous strains it is much easier, since any homogeneous strain corresponds to <u>some</u> tractions, and these need only be measured directly. Empirical tables of $\beth_1$ and $\beth_{-1}$ for certain rubbers have been compiled in this way.

LECTURE 13: SOME NON-HOMOGENEOUS DEFORMATIONS OF ISOTROPIC INCOMPRESSIBLE BODIES.

Universal Deformations in Infinitesimal Elasticity.

The infinitesimal theory of elasticity for static deformations subject to boundary tractions alone is governed by NAVIER's equation:

$$\text{grad div } \underset{\sim}{u} + \alpha\Delta\underset{\sim}{u} = \underset{\sim}{0} , \tag{13.1}$$

where $\underset{\sim}{u}$ is the displacement vector and α is a material coefficient. The solution $\underset{\sim}{u}$ of a particular boundary-value problem generally depends upon α:

$$\underset{\sim}{u} = \underset{\sim}{u}(\underset{\sim}{x},\alpha) . \tag{13.2}$$

That is, the displacement corresponding to the given boundary values for one material is generally different at interior points from that for another. In order that a displacement field $\underset{\sim}{u}(\underset{\sim}{x})$ be possible for <u>all</u> materials, whatever the value of α, by (13.1) it is necessary and sufficient that

$$\begin{aligned} \text{grad div } \underset{\sim}{u} &= \underset{\sim}{0} , \\ \Delta\underset{\sim}{u} &= \underset{\sim}{0} . \end{aligned} \tag{13.3}$$

The three components of $\underset{\sim}{u}$ must then satisfy six second-order partial differential equations.

Physicists call systems whose number is greater than the number of unknowns "overdetermined". According to the religion of physics, such systems have no solutions, because it takes exactly n conditions to determine n quantities. The physicists often enough get on quite well with this principle, which they invoke as a prayer. Its efficiency may be compared with that of the prayers to God by a devout thief, begging that he not be caught.

Most thieves are not caught, but some are. A rational observer is likely to conclude that whether or not a particular thief gets caught has little to do with such prayers as he may offer. The same is true of physicists. Most of the things they believe are more or less true, irrespective of the flimsy grounds they have for believing them. When one of them is not true, the usual physicists' prayer is of no help. In the present case, the faith that n conditions always determine n quantities happens to be unfounded[1], for there are many solutions beyond the trivial rigid one, $\underset{\sim}{u} = \underset{\sim}{0}$. For example, all the St. Venant torsion solutions satisfy the "overdetermined" system (13.3). The corresponding displacements are thus universal solutions in the sense that we need not know α in order to find them.

---

[1] The standard counter-example to the physicists' belief is as follows. We are to determine n real numbers $x_1, x_2, \ldots, x_n$. Clearly the following n conditions suffice:

$$x_1 = 0, \quad x_2 = 0, \ldots, x_n = 0.$$

These are equivalent, however, to 1 condition:

$$x_1^2 + x_2^2 + \ldots + x_n^2 \leqq 0;$$

to 2 conditions:

$$x_1^2 + x_2^2 + \ldots + x_{n-1}^2 \leqq 0, \quad x_n^2 \leqq 0;$$

... to n-1 conditions:

$$x_1^2 + x_2^2 \leqq 0, \quad x_3^2 \leqq 0, \ldots, x_n^2 \leqq 0;$$

to $n + 1$ conditions :

$$x_1(x_1-1) = 0, \quad x_1(x_1-2) = 0, \quad x_2 = 0, \quad x_3 = 0, \ldots, x_n = 0,$$

etc., and the conditions listed in each of the above lines are independent in the sense that no one is determined by the rest, and no proper subset suffices. Thus the "number of conditions" is a completely mystic concept. The counter-example is dismissed by the "physical" because "it does not arise in a physical problem."

Universal solutions may be defined in any field theory. They are the most valuable of all solutions, since they lend themselves easily to determination of parameters such as α . We know they can be produced; we have only to find tractions which will produce them. With a solution that is not universal, the problem is more difficult, since α must be known first in order to determine $\mathbf{u}$ by solving a boundary-value problem.

Universal Deformation Histories in Simple Materials.

The results in Lecture 5 may be interpreted as stating that in any unconstrained homogeneous simple body, all accelerationless homogeneous deformation histories are universal solutions if $\mathbf{b} = \mathbf{0}$, while in homogeneous incompressible bodies, all circulation-preserving homogeneous deformation histories are universal if $\mathbf{b}$ is conservative.

Universal Static Deformations in Unconstrained Elastic Bodies.

It follows from the above-stated result that any homogeneous static deformation is a universal solution for unconstrained homogeneous elastic bodies. It is natural to ask if there be any others. For an isotropic body referred to as an undistorted state, by (11.37) the equation of equilibrium when $\mathbf{b} = \mathbf{0}$ is

$$\operatorname{div}(\beth_0\mathbf{1} + \beth_1\mathbf{B} + \beth_{-1}\mathbf{B}^{-1}) = \mathbf{0} \ . \tag{13.4}$$

For a universal solution, $\mathbf{B}$ must be such as to satisfy this equation identically in the functions $\beth_\Gamma$. If we carry out the differentiations and set the coefficients of $\beth_0, \partial_I\beth_0, \partial_{II}\beth_0, \dots, \partial_{III}\beth_{-1}$ equal to 0, we obtain 12 conditions to be satisfied by the 6 components of $\mathbf{B}$. In addition, $\mathbf{B}$ must be positive-definite and derivable from a deformation field by the formula $\mathbf{B} = \mathbf{F}\mathbf{F}^T$ where $\mathbf{F} = \nabla\boldsymbol{\chi}$. Still more conditions result. ERICKSEN proved that $\mathbf{B}$ = constant is the only tensor satisfying all these conditions. Thus, in this case, the physicists' faith is justified.

Universal Static Deformations In Incompressible Elastic Bodies.

From the theorem in Lecture 5 we know that any homogeneous isochoric deformation is a universal solution for homogeneous incompressible bodies. If we seek more, for isotropic bodies we obtain by (11.45) the following condition:

$$-\text{grad } p + \text{div}(\beth_1 \underset{\sim}{B} + \beth_{-1} \underset{\sim}{B}^{-1}) = \underset{\sim}{0} \ . \quad (13.5)$$

Hence

$$\text{curl div}(\beth_1 \underset{\sim}{B} + \beth_{-1} \underset{\sim}{B}^{-1}) = \underset{\sim}{0} \ . \quad (13.6)$$

If we write out this equation and annul the coefficients of $\beth_1$, $\partial_I \beth_1$, $\partial_I^2 \beth_1$, ... , $\partial_{II}^2 \beth_{-1}$, we obtain 12 conditions again, and now also $\underset{\sim}{B}$ must satisfy also $\det \underset{\sim}{B} = \underset{\sim}{1}$. The physicists' luck runs out at the number 13. Beside $\underset{\sim}{B}$ = const., there are many solutions of (13.6), and their discovery, begun by RIVLIN in the late 1940's, opened the great revival of finite elasticity.

List of the Universal Solutions.

Once a particular solution is known, it is a simple matter to verify it. Below are listed four families of solutions, each depending on several constants A,B,C, *etc*. In this list, capital letters denote co-ordinates in the undistorted reference configuration: X,Y,Z are rectangular Cartesian; R,Θ,Z, cylindrical polar; R,Θ,Φ, spherical polar. Small letters denote co-ordinates in the deformed configuration: x, y, z; r θ, z; and r, θ, ϕ, with standard meanings. In each case the list gives the mapping $\underset{\sim}{x} = \chi(\underset{\sim}{X})$ in terms of components with respect to the co-ordinate systems indicated.

Family 1: Bending, stretching, and shearing of a rectangular block.

$$r = \sqrt{2AX} \ , \quad \theta = BY \ , \quad z = \frac{Z}{AB} - BCY \ , \quad AB \neq 0 \ . \quad (13.7)$$

Family 2: Straightening, stretching, and shearing of a sector of a circular-cylindrical tube.

$$x = \frac{1}{2} AB^2R^2 , \ y = \frac{\Theta}{AB} \ , \quad z = \frac{Z}{B} + \frac{C\Theta}{AB} \ , \quad AB \neq 0 \ . \quad (13.8)$$

Family 3: Inflation or eversion, bending, torsion, extension, and shear of a sector of a circular-cylindrical tube.

$$r = \sqrt{AR^2 + B} \ , \quad \theta = C\Theta + DZ \ , \quad z = E\Theta + FZ \ ,$$
$$A(CF - DE) = 1 \ . \quad (13.9)$$

Family 4: Inflation or eversion of a sector of a spherical shell.

$$r = (\pm R^3 + A)^{\frac{1}{3}} , \quad \theta = \pm \Theta , \quad \phi = \Phi . \tag{13.10}$$

These families are important because the stress systems needed to produce them illustrate the interaction of different kinds of deformations. In the infinitesimal theory, the stress corresponding to two displacements together is the sum of the stresses required to produce each separately. In finite deformation, of course, superposition fails. These families of universal solutions allow us to understand just how it fails in certain cases, as we see below.

Inflation or Eversion, Torsion, and Extension of a Cylinder.

Family 3 is certainly the most interesting. As an illustration of method, we shall work out the details for the most important special case included, namely, that when there is no angular shear $(C = 1)$ and no shear of the generators of the cylinders $(E = 0)$:

$$r = \sqrt{AR^2 + B} , \quad \theta = \Theta + DZ , \quad z = FZ , \quad AF = 1 . \tag{13.11}$$

First we calculate the contravariant components of $\underset{\sim}{B}$, recalling that $B^{km} = x^k{}_{,\alpha} \, x^m{}_{,\beta} \, g^{\alpha\beta}$.

$$\left\| B^{km} \right\| = \left\| \begin{matrix} \frac{A^2R^2}{r^2} & 0 & 0 \\ 0 & \frac{1}{R^2} + D^2 & DF \\ 0 & DF & F^2 \end{matrix} \right\| . \tag{13.12}$$

To calculate $(B^{-1})_{km}$, we may invert the matrix (13.12), or we may use the formula $(B^{-1})_{km} = X^{\alpha}{}_{,k} X^{\beta}{}_{,m} g_{\alpha\beta}$.

$$\left\| (B^{-1})_{km} \right\| = \left\| \begin{matrix} \frac{r^2}{A^2R^2} & 0 & 0 \\ 0 & R^2 & -ADR^2 \\ 0 & -ADR^2 & A\,(1+D^2R^2) \end{matrix} \right\| , \tag{13.13}$$

$$I = \operatorname{tr} \underset{\sim}{B} = g_{km} B^{km} = \frac{A^2 R^2}{r^2} + r^2 \left(\frac{1}{R^2} + D^2\right) + F^2 \ ,$$

$$II = \operatorname{tr} \underset{\sim}{B}^{-1} = g^{km} (B^{-1})_{km} = \frac{r^2}{A^2 R^2} + \frac{R^2}{r^2} + A^2(1 + D^2 R^2) \ .$$

The components of the determinate stress $\underset{\sim}{T} + p\underset{\sim}{1}$ follow by putting (13.12) and (13.13) into (11.45).

For the moment, not needing the full results, we remark only on three facts concerning them:

$$\underset{\sim}{T} + p\underset{\sim}{1} = \underset{\sim}{f}(r) \ ,$$

$$T\langle r\theta\rangle = T\langle rz\rangle = 0 \ . \tag{13.14}$$

We proceed to show in any case when (13.14) holds, it is possible to choose p in such a way as to render the stress system $\underset{\sim}{T}$ compatible with Cauchy's first law (3.16) when $\underset{\sim}{b} = \underset{\sim}{0}$. This is so because in cylindrical co-ordinates that law then assumes the form

$$\partial_r T\langle rr\rangle + \frac{1}{r}\left(T\langle rr\rangle - T\langle\theta\theta\rangle\right) = 0 \ ,$$

$$\partial_\theta p = 0 \ , \tag{13.15}$$

$$\partial_z p = 0 \ .$$

Hence $p = p(r)$, and

$$T\langle rr\rangle = -\int \frac{T\langle rr\rangle - T\langle\theta\theta\rangle}{r}\, dr \ . \tag{13.16}$$

Conversely, if (13.16) is satisfied and $p = p(r)$, the stress system is equilibrated, subject to boundary tractions alone. It should be noticed that (13.16) amounts to a determination of $p(r)$ when $\underset{\sim}{f}$ in $(13.14)_1$ is known. However, we prefer to regard it as eliminating p altogether from the problem. An alternative form of (13.16) is

$$T\langle\theta\theta\rangle = (r\, T\langle rr\rangle)' \ , \tag{13.17}$$

since $\underset{\sim}{T}$ has now been shown to be a function of r only.

<u>Exercise 13.1</u>. Prove that the resultant normal traction N on the part of the plane end $z = \text{const.}$ bounded by the circles $r = r_1$ and $r = r_2$ is

$$N = \pi r^2 T\langle rr\rangle \Big|_{r_1}^{r_2} + \pi \int_{r_1}^{r_2} (2T\langle zz\rangle - T\langle rr\rangle - T\langle\theta\theta\rangle) r\,dr \,. \tag{13.18}$$

Choice of the constant of integration in (13.16) allows us to render any one cylinder, say $r = r_1$ or $r = r_2$, free of traction, but generally not more than one. By (13.18), it is not generally possible to produce the deformation without supplying definite normal tensions on the planes $z = \text{const.}$, and these tensions generally have a non-zero resultant N.

The full solution of the problem is now obtained by substituting (13.12) and (13.13) into (11.45). The shear stress is given directly:

$$T\langle\theta z\rangle = DFR^2\Big(\frac{r}{R^2}\beth_1 - \frac{A^2}{r}\beth_{-1}\Big) \,. \tag{13.19}$$

Explicit formulae result also for, say, $T\langle rr\rangle - T\langle\theta\theta\rangle$ and for $T\langle zz\rangle - T\langle rr\rangle$. To calculate $T\langle rr\rangle$, we substitute the former in the right-hand side of (13.16).

The results will be illustrated now in major special cases.

<u>Torsion and Tension of a Solid Cylinder</u>.

Set $B = 0$, $r_1 = R_1 = 0$ in (13.11), and choose the constant of integration in (13.16) so that $T\langle rr\rangle = 0$ when $R = R_2$. The deformation then represents a twist D/F superimposed upon a longitudinal stretch F in a solid cylinder with free mantle. The resultant torque T and resultant normal tension N on the plane ends are given by (13.19) and (13.18):

$$\begin{aligned} \frac{T}{D} &= \frac{1}{D}\int_0^{r_2} rT\langle z\theta\rangle \cdot 2\pi r\,dr \,, \\ &= \frac{2\pi}{F}\int_0^{R_2} \Big(\beth_1 - \frac{1}{F}\beth_{-1}\Big) R^3\,dR \,, \end{aligned} \tag{13.20}$$

$$N = 2\pi\left(F - \frac{1}{F^2}\right)\int_0^{R_2}\left(\beth_1 - \frac{1}{F}\beth_{-1}\right)R\,dR - \frac{\pi D^2}{F}\int_0^{R_2}\left(\beth_1 - \frac{2}{F}\beth_{-1}\right)R^2\,dR. \qquad (13.20)$$

Since the arguments of $\beth_1$ and $\beth_{-1}$ are I and II as given by (13.13), T and N are both sums of two functions of F and D which are not functions of either variable alone. True, the first integral in the expression for N vanishes if the extension vanishes (F = 1), and the second one vanishes if the twist vanishes (D = 0), but each is a function of both F and D in general. Thus it is impossible to separate the normal tension into a part "due" to the twist and another "due" to the stretch.

Three major problems are solved by (13.20). In <u>Coulomb's problem</u> we seek the relation between torque and twist for small twist. Thus we hold F fixed and let D approach 0. If we set

$$\tilde{\beth}_r(F) \equiv \beth_r\left(F + \frac{2}{F},\ \frac{1}{F^2} + 2F\right), \qquad (13.21)$$

then

$$\begin{aligned} \tau(F) &\equiv \lim_{D\to 0} \frac{T}{D/F} = \frac{1}{2}\pi R_2^4\left(\tilde{\beth}_1 - \frac{1}{F}\tilde{\beth}_{-1}\right), \\ N_0(F) &\equiv \lim_{D\to 0} N = \pi R_2^2\left(F - \frac{1}{F^2}\right)\left(\tilde{\beth}_1 - \frac{1}{F}\tilde{\beth}_{-1}\right). \end{aligned} \qquad (13.22)$$

N_0 is the resultant tension needed to produce the stretch F when there is no twist, while τ is the torsional modulus at the extension corresponding to N_0. Both are determined as rather simple functions of F when the response coefficients $\beth_r$ are known. One important result may be obtained without knowing $\beth_r$, namely,

$$\frac{R_2^2 N_0(F)}{\tau(F)} = 2\left(F - \frac{1}{F^2}\right). \qquad (13.23)$$

This elegant <u>universal relation</u> of RIVLIN may be regarded as solving COULOMB's problem for a circular cylinder of incompressible isotropic elastic material. If $N_0(F)$ is determined in any way, <u>e.g</u>. empirically, $\tau(F)$ may be calculated. The results of a tensile test enable us to predict the results of a torsion test. It is theorems of this kind, connecting one phenomenon with another, that continuum mechanics prizes above everything else.

In <u>Poynting's problem</u> we seek the elongation that results from torsion of a cylinder with free ends. To obtain the solution from (13.20), hold D fixed, set $N = 0$, and solve for F. It is not known whether a solution exists in general, but for small twist it does.

<u>Exercise 13.2</u>. Prove that

$$\lim_{D\to 0} \frac{F-1}{(D/F)^2} = \frac{1}{12} R_2^2 \left(\frac{\mu - \tilde{\beth}_{-1}(1)}{\mu} \right) . \tag{13.24}$$

This result, also, is of central importance for the theory. First, it shows that the effect of small torsion is to produce extension ultimately proportional to the square of the twist. Second, there have been many attempts, using special and spurious arguments, to calculate the magnitude of the Poynting effect in terms of the concepts of the infinitesimal theory. In that theory, for isotropic incompressible materials, there is but a single elastic modulus, namely, μ. The exact and general result (13.24) shows that any such attempt is certain to fail, for <u>two</u> moduli, μ and $\beth_{-1}(1)$, are required. Thus it is impossible to describe the Poynting effect properly without information from <u>outside</u> the infinitesimal theory. Third, (13.24) predicts that torsion of a solid incompressible isotropic elastic cylinder results in

$$\begin{matrix} \text{elongation} \\ \\ \text{contraction} \end{matrix} \quad \text{if} \quad \frac{\tilde{\beth}_{-1}(1)}{\mu} \begin{matrix} < \\ \\ > \end{matrix} 1 . \tag{13.25}$$

Experiments on homogeneous strains of rubber sheets give values of $\beth_{-1}(I,II)$ which are negative for all values of I and II. We expect, then, that cylinders of these same rubbers will <u>always elongate</u> in torsion, and they do, as POYNTING observed in 1913.

In some books for engineers may be found an argument, deriving from YOUNG and MAXWELL, which shows that if a wire is idealized as being a bundle of perfectly slippery inextensible rods, it will always shorten when twisted. Neither YOUNG nor MAXWELL claimed this model to be a good one, but the textbooks, always drawn to a bad "intuitive" theory even if contradicted by experiment, continue to present it. If a rod is idealized as a bundle of wires, those wires are not perfectly smooth; they exert both normal and tangential tractions upon one another, and the only known way to calculate the effects of such unknown tractions is to forget about the wires and solve the problem by real elasticity theory. The result, for incompressible materials, is (13.24). It shows that there is no a priori reason to expect one sign or the other, but it enables us to correlate two classes of experiments. The data taken empirically on homogeneous strains enable us to predict lengthening, not only in quality but in amount. Again, it is results of this type that continuum mechanics is designed to deliver.

Finally, we consider pure torsion by setting $F = 1$. Comparison of (13.19) with (12.7) now yields

$$T\langle\theta z\rangle = Dr\,\hat{\mu}(D^2r^2) \quad , \qquad (13.26)$$

where $\hat{\mu}(k^2)$ is the generalized shear modulus. This result confirms the common intuitive claim that torsion is "equivalent" or "analogous" to shear, since the shear stress in torsion is determined by the shear modulus alone. Similarly, the torsional modulus determines the resultant torque by $(13.20)_2$. While "intuition" has been confirmed, its triumph stops at this point. In an unconstrained material the shear modulus is well defined because simple shear is a universal solution. In view of ERICKSEN's theorem, stated at the beginning of this lecture, torsion is not a universal solution for unconstrained materials. A fortiori, response in torsion cannot be determined by any quantity defined by a universal solution. In particular, it cannot be determined by the shear modulus.

Eversion.

A natural problem for study in elasticity is furnished by eversion. If an infinitely long hollow cylinder is turned inside out, it will assume some everted shape, presumably again a circular cylinder, subject to vanishing surface tractions on its mantles. The problem, then, for a tube of inner and outer radii R_1 and R_2, where $R_1 < R_2$, is to determine the possible radii r_1 and r_2 on which the traction vanishes. The trivial solution is $r_1 = R_1$, $r_2 = R_2$. If a second solution exists in which $r_1 > r_2$, it corresponds to eversion. To solve this problem we return to (13.11) and set $D = 0$, leaving A and B to be adjusted. We choose the constant of integration in (13.16) so that $T\langle rr\rangle = 0$ when $R = R_1$. The condition that

$T\langle rr\rangle = 0$ when $R = R_2$ is

$$0 = \int_{R_1}^{R_2} \left[\frac{R^2}{(R^2 + B/A)^2} - \frac{1}{R^2}\right]\left[\beth_1 - \frac{1}{A^2}\beth_{-1}\right] R\,dR = 0 \,. \tag{13.27}$$

A second equation is obtained by setting $N = 0$ in the appropriate special case of (13.18).

Exercise 13.3. Prove that the condition $N = 0$ takes the form

$$\begin{aligned} 0 = \int_{R_1}^{R_2} \Big[&\Big(\frac{A^2R^2}{AR^2 + B} - \frac{2}{A^2} + \frac{AR^2 + B}{R^2}\Big)\beth_1 \\ &+ \Big(\frac{AR^2 + B}{A^2R^2} - 2A + \frac{R^2}{AR^2 + B}\Big)\beth_{-1}\Big] R\,dR \,. \end{aligned} \tag{13.28}$$

A and B are to be found by solving (13.27) and (13.28). If a solution exists in which $A < 0$, it represents eversion. The corresponding stretch is given by $F = 1/A$. In special cases, such a solution is known to exist.

Exercise 13.4. In the special case when $\beth_1$ is a positive constant and $\beth_{-1} = 0$, calculate F explicitly.

Remarks on Method.

Most of the solutions listed were first found by a method of trial. Once one has an idea that a certain deformation should be a universal solution, one has only to try it out. When I first saw some of these solutions, I proposed to my class in elasticity in the spring of 1951 the problem of the inflation of a sphere. Four or five of the students found that (13.10) is indeed a universal solution. Just at that time appeared a paper by GREEN & SHIELD, written about a year earlier, in which that solution was verified and interpreted.

One of the students in that class was J.L. ERICKSEN. Shortly afterward he began to work on the problem of finding all universal solutions. He published his analysis of the problem, a difficult piece of work, in 1954. In the course of it, he discovered Family 2. His results show that beyond homogeneous strain and the four families listed, there are only two possibilities, highly overdetermined, for further identical solutions. I did not read his paper so carefully as I ought have, and in my Dallas lectures of 1960 I claimed that he had proved further universal solutions impossible. Although in several later publications I took pains to correct this misimpression, my Dallas lectures prevailed, and I am widely quoted as the author of this mistake. While the explanation of the greater currency of my Dallas lectures may lie in the fact that the volume was given away for the asking, I am inclined to attribute it to a general principle by which one mistake draws more attention to its author than do a dozen truths.

In any case, until quite recently it was commonly conjectured that only proof was lacking to fill the gaps, and that no more universal solutions could be found. Long unproved conjectures, however, often turn out to be false, and last spring two students at Brown, M. KLINGBEIL and M. SYNGH, working with SHIELD and PIPKIN, respectively, found a new universal solution:

$$r = AR \,, \quad \theta = B \log R + C\Theta \,, \quad z = Z/(A^2C) \,. \tag{13.29}$$

An easy calculation shows that the physical components of $\underset{\sim}{B}$ for this deformation are constants, so it furnishes an example of a stress which is homogeneous and hence equilibrated by boundary tractions, although the deformation itself is not homogeneous. That is, if $\underset{\sim}{F} = \underset{\sim}{R}\underset{\sim}{U}$, $\underset{\sim}{U}$ is homogeneous, but $\underset{\sim}{R}$ is not.

LECTURE 14. INFINITESIMAL DEFORMATION

Displacement Vector and Displacement Gradient.

The displacement vector $\underset{\sim}{u}$ is the vector that transforms the place $\underset{\sim}{X}$ in $\underset{\sim}{\kappa}$ into the place $\underset{\sim}{x}$ in $\underset{\sim}{\chi}$:

$$\underset{\sim}{u} \equiv \underset{\sim}{x} - \underset{\sim}{X} = \underset{\sim}{\chi}(\underset{\sim}{X}, t) - \underset{\sim}{X}, \tag{14.1}$$

Its gradient $\underset{\sim}{H}$ is called the displacement gradient:

$$\underset{\sim}{H} \equiv \nabla \underset{\sim}{u} = \underset{\sim}{F} - \underset{\sim}{1}. \tag{14.2}$$

To introduce the displacement is useful in continuum mechanics only when the displacement and its gradient are small in some sense.

Infinitesimal Displacement.

Corresponding to any given displacement field $\underset{\sim}{u}(\underset{\sim}{X}, t)$, consider the family of displacements $\varepsilon\underset{\sim}{u}$ as $\varepsilon \to 0$, and denote quantities defined from it with a subscript ε. Thus

$$\underset{\sim}{H}_\varepsilon \equiv \nabla(\varepsilon\underset{\sim}{u}) = \varepsilon\underset{\sim}{H}. \tag{14.3}$$

Then

$$\begin{aligned} \underset{\sim}{U}_\varepsilon^2 = \underset{\sim}{F}_\varepsilon^T\underset{\sim}{F}_\varepsilon &= (\underset{\sim}{1} + \varepsilon\underset{\sim}{H})^T(\underset{\sim}{1} + \varepsilon\underset{\sim}{H}), \\ &= \underset{\sim}{1} + \varepsilon(\underset{\sim}{H} + \underset{\sim}{H}^T) + O(\varepsilon^2). \end{aligned} \tag{14.4}$$

Thus

$$\underset{\sim}{U}_\varepsilon = \underset{\sim}{1} + \frac{1}{2}\varepsilon(\underset{\sim}{H} + \underset{\sim}{H}^T) + O(\varepsilon^2). \tag{14.5}$$

Since the polar decomposition of $\underset{\sim}{F}_\varepsilon$ is $\underset{\sim}{F}_\varepsilon = \underset{\sim}{R}_\varepsilon \underset{\sim}{U}_\varepsilon$, it is apparent from (14.5) that

$$\underset{\sim}{R}_\varepsilon = \underset{\sim}{1} + \frac{1}{2}\,\varepsilon(\underset{\sim}{H} - \underset{\sim}{H}^T) + O(\varepsilon^2). \tag{14.6}$$

Thus if we set

$$\tilde{\underset{\sim}{E}} \equiv \frac{1}{2}\,(\underset{\sim}{H} + \underset{\sim}{H}^T), \qquad \tilde{\underset{\sim}{R}} \equiv \frac{1}{2}\,(\underset{\sim}{H} - \underset{\sim}{H}^T), \tag{14.7}$$

we have to within terms $O(\varepsilon^2)$

$$\begin{aligned} \underset{\sim}{U}_\varepsilon &= \underset{\sim}{1} + \varepsilon\tilde{\underset{\sim}{E}} = \underset{\sim}{V}_\varepsilon, \\ \underset{\sim}{R}_\varepsilon &= \underset{\sim}{1} + \varepsilon\tilde{\underset{\sim}{R}}, \\ \underset{\sim}{R}_\varepsilon^T &= \underset{\sim}{1} - \varepsilon\tilde{\underset{\sim}{R}}. \end{aligned} \tag{14.8}$$

$\tilde{\underset{\sim}{E}}$ and $\tilde{\underset{\sim}{R}}$ are called the tensors of infinitesimal strain and rotation. The results (14.8) show that when ε is sufficiently small, the exact polar decompositions $\underset{\sim}{F}_\varepsilon = \underset{\sim}{R}_\varepsilon \underset{\sim}{U}_\varepsilon = \underset{\sim}{V}_\varepsilon \underset{\sim}{R}_\varepsilon$ are replaced by the additive resolution of $\underset{\sim}{H}_\varepsilon$ into its symmetric and skew parts: $\underset{\sim}{H}_\varepsilon = \tilde{\underset{\sim}{E}}_\varepsilon + \tilde{\underset{\sim}{R}}_\varepsilon$. A theory in which only the terms of lowest non-trivial order in ε are retained is described as a theory of infinitesimal deformations.

Constitutive Equation for Infinitesimal Elastic Deformations.

Consider the reduced form (11.6) for the constitutive equation of an elastic material:

$$\underset{\sim}{T} = \underset{\sim}{R}\, \underset{\sim}{\mathcal{G}}(\underset{\sim}{U})\underset{\sim}{R}^T. \tag{11.6}$$

and assume that $\underset{\sim}{\mathcal{G}}$ is continuously differentiable at $\underset{\sim}{U} = \underset{\sim}{1}$. For the family of displacements $\varepsilon\underset{\sim}{u}$, by $(14.8)_1$

$$\begin{aligned} \underset{\sim}{\mathcal{G}}(\underset{\sim}{U}) &= \underset{\sim}{\mathcal{G}}(\underset{\sim}{1} + \varepsilon\tilde{\underset{\sim}{E}} + O(\varepsilon^2)) \\ &= \underset{\sim}{T}_0 + \varepsilon L[\tilde{\underset{\sim}{E}}] + O(\varepsilon^2), \end{aligned} \tag{14.9}$$

where L is a linear mapping of symmetric tensors into symmetric tensors, and where $\underset{\sim}{T}_0 = \underset{\sim}{\mathcal{G}}(\underset{\sim}{1})$, the stress tensor in the body if it actually occupies the reference configuration. Explicitly,

$$L = \partial_{\underset{\sim}{U}} \underset{\sim}{\mathcal{G}}(\underset{\sim}{U}) \Big|_{\underset{\sim}{U} = \underset{\sim}{1}} . \tag{14.10}$$

Substitution of (14.9) and $(14.8)_2$ into (11.6) yields

$$\begin{aligned} \underset{\sim}{T} &= (\underset{\sim}{1} + \varepsilon\underset{\sim}{\tilde{R}})(\underset{\sim}{T}_0 + \varepsilon L[\underset{\sim}{\tilde{E}}])(\underset{\sim}{1} - \varepsilon\underset{\sim}{\tilde{R}}) + O(\varepsilon^2) \\ &= \underset{\sim}{T}_0 + \varepsilon(\underset{\sim}{\tilde{R}}\underset{\sim}{T}_0 - \underset{\sim}{T}_0\underset{\sim}{\tilde{R}} + L[\underset{\sim}{\tilde{E}}]) + O(\varepsilon^2). \end{aligned} \tag{14.11}$$

The foregoing analysis shows that, given any family of displacements $\varepsilon\underset{\sim}{u}$ in an elastic material whose response function $\underset{\sim}{\mathcal{G}}(\underset{\sim}{F})$ is continuously differentiable at $\underset{\sim}{F} = \underset{\sim}{1}$, the stress is sufficiently approximated by (14.11) if ε is made small enough. Accordingly, the <u>constitutive equation for infinitesimal displacements</u> is given by <u>CAUCHY's law</u>:

$$\underset{\sim}{T} = \underset{\sim}{T}_0 + \underset{\sim}{\tilde{R}}\underset{\sim}{T}_0 - \underset{\sim}{T}_0\underset{\sim}{\tilde{R}} + L[\underset{\sim}{\tilde{E}}], \tag{14.12}$$

where $\underset{\sim}{T}_0$ is the stress in the configuration from which the infinitesimal displacements are reckoned.

<u>Exercise 14.1</u>. Prove the <u>POISSON-CAUCHY theorem</u> (1829): The constitutive equation

$$\underset{\sim}{T} = \underset{\sim}{f}(\underset{\sim}{H}), \tag{14.13}$$

where $\underset{\sim}{f}$ is linear and $\underset{\sim}{H}$ is limited to infinitesimal deformations, is frame-indifferent if and only if it reduces to (14.12). (Here (11.6) is not laid down as the starting point, so (14.10) does not necessarily follow, nor need the initial stress $\underset{\sim}{T}_0$ correspond to elastic deformation from some other configuration.)

The fourth-order tensor L is called the <u>linear elasticity</u> of the elastic material for the configuration $\underset{\sim}{\kappa}$. Of course, the same material has different linear elasticities for different configurations $\underset{\sim}{\kappa}$.

<u>Classical Infinitesimal Elasticity</u>.

In the classical theory it is assumed that the elastic material has a natural state, and only infinitesimal deformations from that state are considered. Then (14.12) reduces to

$$\underset{\sim}{T} = L[\underset{\sim}{\tilde{E}}]. \tag{14.14}$$

When the body is isotropic and the natural state is undistorted, (14.14) assumes the form

$$\underset{\sim}{T} = \lambda(\mathrm{tr}\ \underset{\sim}{\tilde{E}})\underset{\sim}{1} + 2\mu\underset{\sim}{\tilde{E}}, \tag{14.15}$$

as may be concluded by specialization from (11.37), by direct application of (11.27), or by a simple independent argument. Many of the developments given in books on the classical theory can be easily extended to infinitesimal deformations from an arbitrary configuration, with the more general constitutive equation (14.12).

Relation of the Linear Elasticity to the General One.

In many cases of application it is desirable to study the effects of an infinitesimal deformation superimposed upon a configuration $\underset{\sim}{\kappa}$ which is obtained from another configuration $\underset{\sim}{\kappa}_0$ by a known deformation with gradient $\underset{\sim}{F}_0$ and corresponding stress $\underset{\sim}{T}_{\underset{\sim}{\kappa}}$. For example, $\underset{\sim}{\kappa}_0$ might be a homogeneous natural undistorted state of a solid. Then $\underset{\sim}{\kappa}$ would be, in general, an inhomogeneous distorted state. The problem here is to calculate the linear elasticity L with respect to $\underset{\sim}{\kappa}$ in terms of the response function $\underset{\sim}{\mathfrak{g}}_{\underset{\sim}{\kappa}_0}$ with respect to the configuration $\underset{\sim}{\kappa}_0$.

In this case, if we write $\underset{\sim}{F}^*$ for the gradient of the deformation from $\underset{\sim}{\kappa}_0$ to $\underset{\sim}{\chi}$,

$$\underset{\sim}{F}^* = \underset{\sim}{F}\underset{\sim}{F}_0 = (\underset{\sim}{H} + \underset{\sim}{1})\underset{\sim}{F}_0, \tag{14.16}$$

when $\underset{\sim}{F}_0$ is the gradient of the deformation from $\underset{\sim}{\kappa}_0$ to $\underset{\sim}{\kappa}$. If $\underset{\sim}{\kappa}_0$ is used as reference configuration, the Piola-Kirchhoff stress $\underset{\sim}{T}_{\underset{\sim}{\kappa}_0}$ in $\underset{\sim}{\kappa}$ is given by

$$\underset{\sim}{T}_{\underset{\sim}{\kappa}_0}(\underset{\sim}{F}_0) = \underset{\sim}{\mathfrak{h}}_{\underset{\sim}{\kappa}_0}(\underset{\sim}{F}_0), \tag{14.17}$$

so by $(11.13)_2$

$$\mathbf{T} = \frac{\rho_{\kappa}}{\rho_{\kappa_0}} \mathfrak{h}_{\kappa_0}(\mathbf{F}_0)\mathbf{F}_0^T = \mathfrak{h}_{\kappa}(\mathbf{1}),$$

$$\mathbf{T}_{\kappa_0}(\mathbf{F}^*) = \mathfrak{h}_{\kappa_0}[(\mathbf{H} + \mathbf{1})\mathbf{F}_0], \tag{14.18}$$

$$= \mathfrak{h}_{\kappa_0}(\mathbf{F}_0) + A_{\kappa_0}(\mathbf{F}_0)[\mathbf{H}\mathbf{F}_0],$$

where A_{κ_0} is the elasticity A defined by $(11.20)_{2,3}$, using κ_0 as reference configuration. Hence

$$\mathbf{T} = \frac{\rho_{\kappa}}{\rho} \mathbf{T}_{\kappa_0}(\mathbf{F}^*)\mathbf{F}^{*T},$$

$$= \frac{\rho_{\kappa_0}}{\rho} \cdot \frac{\rho_{\kappa}}{\rho_{\kappa_0}} (\mathfrak{h}_{\kappa_0}(\mathbf{F}_0) + A_{\kappa_0}(\mathbf{F}_0)[\mathbf{H}\mathbf{F}_0])\mathbf{F}_0^T(\mathbf{H}^T + \mathbf{1}),$$

$$= \frac{\rho_{\kappa_0}}{\rho} (\mathbf{T}_0 + \frac{\rho_{\kappa}}{\rho_{\kappa_0}} A_{\kappa_0}(\mathbf{F}_0)[\mathbf{H}\mathbf{F}_0]\mathbf{F}_0^T)(\mathbf{H}^T + \mathbf{1}), \tag{14.19}$$

$$= (\mathbf{1} + \operatorname{tr} \tilde{\mathbf{E}})\mathbf{T}_0(\mathbf{H}^T + \mathbf{1}) + \frac{\rho_{\kappa}}{\rho_{\kappa_0}} A_{\kappa_0}(\mathbf{F}_0)[\mathbf{H}\mathbf{F}_0]\mathbf{F}_0^T,$$

$$= \mathbf{T}_0 + (\operatorname{tr} \tilde{\mathbf{E}})\mathbf{T}_0 + \mathbf{T}_0\mathbf{H}^T + \frac{\rho_{\kappa}}{\rho_{\kappa_0}} A_{\kappa_0}(\mathbf{F}_0)[\mathbf{H}\mathbf{F}_0]\mathbf{F}_0^T.$$

Comparison with (14.12) shows that

$$L_{\kappa}[\tilde{\mathbf{E}}] = \mathbf{T}_0\tilde{\mathbf{E}} + (\operatorname{tr} \tilde{\mathbf{E}})\mathbf{T}_0 - \tilde{\mathbf{R}}\mathbf{T}_0 + \frac{\rho_{\kappa}}{\rho_{\kappa_0}} A_{\kappa_0}(\mathbf{F}_0)[\mathbf{H}\mathbf{F}_0]\mathbf{F}_0^T \tag{14.20}$$

This formula gives the linear elasticity L_{κ} with respect to κ in terms of the stress $\mathbf{T}_0$ in κ and the elasticity $A_{\kappa_0}(\mathbf{F}_0)$ with respect to κ_0, evaluated in κ.

It is not evident that $L_{\kappa}[\tilde{\mathbf{E}}]$ as delivered by (14.20) is a symmetric tensor, but in fact it is.

Exercise 14.2. Find an identity satisfied by A in virtue of the principle of frame-indifference (expressed, *e.g.*, by the identity (11.16)), and thence prove that the value of the right-hand side of (14.20) is a symmetric tensor.

If κ_0 is a homogeneous configuration of a homogeneous body, clearly κ will generally fail to be such. Explicitly, if the gradient F_0 of the deformation from κ_0 to κ is not homogeneous, the elasticity L_κ as given by (14.20) generally will be a function of X, even if L_{κ_0} is not. If κ_0 is an undistorted state of an elastic solid, clearly κ will generally fail to be so, as also is clear from (14.20). If κ_0 is a natural state, so that (14.14) holds, generally κ is not a natural state, and the stress $T - T_0$ corresponding to infinitesimal deformation from κ depends not only on the strain $\tilde{E}$ but also on the rotation $\tilde{R}$. These simple and evident facts are sometimes described in misleading ways: A severely strained elastic body "loses" its homogeneity, isotropy, etc., or even "is no longer elastic".

If the elastic body is isotropic and the reference configuration κ_0 is undistorted, the linear elasticity L_κ can be calculated in explicit form. While this form may be obtained by specialization from (14.20), it is easier to start afresh from the stress relation (11.37).

Exercise 14.2. Let B be the left Cauchy-Green tensor for the deformation from an undistorted state of an isotropic material to one in which the stress is, say, T_0:

$$T_0 = \beth_0 1 + \beth_1 B + \beth_{-1} B^{-1}. \tag{14.21}$$

Prove that in a further infinitesimal deformation

$$\begin{aligned} T = T_0 + HT_0 + T_0 H^T - 2[\beth_0 \tilde{E} + \beth_{-1}(\tilde{E}B^{-1} + B^{-1}\tilde{E})] \\ + 2\sum_{r=-1}^{+1} \{(II \frac{\partial \beth_r}{\partial II} + III \frac{\partial \beth_r}{\partial III}) \operatorname{tr} \tilde{E} \\ + \frac{\partial \beth_r}{\partial II} \operatorname{tr}(B\tilde{E}) - III \frac{\partial \beth_r}{\partial II} \operatorname{tr}(B^{-1}\tilde{E})\} B^r, \end{aligned} \tag{14.22}$$

where I, II, and III are the principal invariants of B.

The main applications of these results have been to the case when F_0 corresponds to homogeneous deformation of a homogeneous isotropic body from a natural state. Then $\beth_0 + \beth_1 + \beth_{-1} = 0$ when $I = II = 3$ and

III = 1, and $\underset{\sim}{B}$ is a constant. The most important special result obtained is GREEN & SHIELD's general solution of COULOMB's problem for an isotropic body. A cylinder of arbitrary simply-connected cross-section is stretched along its generators and then subjected to infinitesimal twist. The homogeneous strain is then given by

$$x_1 = \alpha v X_1, \quad x_2 = \alpha v X_2, \quad x_3 = v X_3, \tag{14.23}$$

where the stretch v along the X_3-direction is assigned and where α is determined by the condition that the transverse tractions shall vanish, so that the mantle of the cylinder is free. The superimposed small torsion is assumed to be of ST. VENANT's type:

$$u_1 = -\varepsilon x_2 x_3, \quad u_2 = \varepsilon x_1 x_3, \quad u_3 = \varepsilon \Phi(x_1, x_2), \tag{14.24}$$

where ε, the twist, is small.

<u>Exercise 14.3</u>. Prove that a solution of the form (14.23) is possible if and only if

$$\Phi(x_1, x_2) = \alpha^2 v^2 \phi\left(\frac{x_1}{\alpha v}, \frac{x_2}{\alpha v}\right), \tag{14.25}$$

where ϕ is ST. VENANT's warping function for the given cross-section.

The result (14.25) reduces the solution of COULOMB's problem to the solution of ST. VENANT's torsion problem in the classical theory. The calculations are long but routine. The torsional modulus $\tau(v)$ turns out to have the following form:

$$\tau(v) = \frac{M}{\varepsilon} = \alpha^4 v^4 (v^2 \tilde{\beth}_1 - \alpha^{-2} v^{-2} \tilde{\beth}_{-1})[I_0 - \alpha^2 (I_0 - S_0)], \tag{14.26}$$

where S_0 is the torsional rigidity of the cross-section for the natural state and where I_0 is the polar moment of inertia of the cross-section with respect to its centroid. The resultant normal traction N is related to $\tau(v)$ through the identity

$$\frac{N}{\tau(v)} = \frac{1 - \alpha^2}{\alpha^2 v^2} \cdot \frac{A_0}{I_0 - \alpha^2 (I_0 - S_0)}, \tag{14.27}$$

where A_0 is the cross-sectional area. The subscripts "0" indicate that the quantities to which they are attached are those for the natural state, before the stretch v is effected. The results (14.26) and (14.27) are to be compared with their counterparts (13.22) and (13.23) for the incompressible cylinder of circular cross-section.

The general nature of α as a function of v is not known. If we assume, as seems plausible, that a compressed cylinder thickens, then $\alpha > 1$ if $v < 1$. It is known that $S_0 < I_0$ if the cross-section is not a circular disk. Then, according to (14.27), $\tau(v) = 0$ when $\alpha(v)$ reaches a value such that

$$\frac{1}{\alpha^2} = \frac{I_0 - S_0}{I_0} . \tag{14.28}$$

That is, sufficient compression destroys the torsional strength of the cylinder.

LECTURE 15. ITERATIVE SOLUTION OF THE BOUNDARY-VALUE PROBLEM OF TRACTION

Formulation of the Method.

The routine approach to non-linear problems is through a process of perturbation. In elasticity a method of series expansion was introduced by SIGNORINI along standard lines, for the boundary-value problem of traction.

The stress relation of elasticity, hitherto used in the forms (11.2) and (11.15), is now written in the equivalent form

$$\underset{\sim}{T}_{\underset{\sim}{\kappa}} = \underset{\sim}{\mathfrak{K}}(\underset{\sim}{H}), \tag{15.1}$$

where $\underset{\sim}{H}$ is the displacement gradient:

$$\underset{\sim}{H} = \nabla\underset{\sim}{u} = \underset{\sim}{F} - \underset{\sim}{1}. \tag{14.2}$$

It is now assumed that $\underset{\sim}{\mathfrak{K}}$ is an analytic function of $\underset{\sim}{H}$ and that the reference configuration is a natural state:

$$\underset{\sim}{T}_{\underset{\sim}{\kappa}} = \underset{\sim}{\mathfrak{K}}(\underset{\sim}{H}) = \sum_{s=1}^{\infty} \underset{\sim}{\mathfrak{K}}_s(\underset{\sim}{H}), \tag{15.2}$$

where $\underset{\sim}{\mathfrak{K}}_s$ is a homogeneous polynomial of degree s. It is assumed also that the body force $\underset{\sim}{b}$ in $\mathcal{B}_{\underset{\sim}{\kappa}}$ and traction $\underset{\sim}{t}_{\underset{\sim}{\kappa}}$ on $\partial\mathcal{B}_{\underset{\sim}{\kappa}}$ are analytic functions of a parameter ε:

$$\begin{aligned} \underset{\sim}{b} &= \sum_{n=1}^{\infty} \varepsilon^n \underset{\sim}{b}_n \quad \text{in } \mathcal{B}_{\underset{\sim}{\kappa}}, \\ \underset{\sim}{t}_{\underset{\sim}{\kappa}} &= \sum_{n=1}^{\infty} \varepsilon^n \underset{\sim}{t}_{\underset{\sim}{\kappa} n} \quad \text{on } \partial\mathcal{B}_{\underset{\sim}{\kappa}}. \end{aligned} \tag{15.3}$$

We seek a solution of the boundary-value problem of traction with the data (15.3) which is an analytic function of ε:

$$\underset{\sim}{u} = \sum_{n=1}^{\infty} \varepsilon^n \underset{\sim}{u}_n . \tag{15.4}$$

If there is such a solution, then

$$\underset{\sim}{H} = \sum_{n=1}^{\infty} \varepsilon^n \underset{\sim}{H}_n, \text{ where } \underset{\sim}{H}_n = \nabla \underset{\sim}{u}_n, \tag{15.5}$$

By (15.3), then,

$$\begin{aligned} \underset{\sim}{T}_{\underset{\sim}{\kappa}} &= \sum_{s=1}^{\infty} \underset{\sim}{\mathfrak{K}}_s \left(\sum_{r=1}^{\infty} \varepsilon^r \underset{\sim}{H}_r \right) \\ &= \sum_{n=1}^{\infty} \varepsilon^n \underset{\sim}{T}_{\underset{\sim}{\kappa} n}(\underset{\sim}{H}_1, \underset{\sim}{H}_2, \ldots, \underset{\sim}{H}_n), \end{aligned} \tag{15.6}$$

where the function $\underset{\sim}{T}_{\underset{\sim}{\kappa} n}$ may be calculated explicitly in terms of the functions $\underset{\sim}{\mathfrak{K}}_s$. For example,

$$\underset{\sim}{T}_{\underset{\sim}{\kappa} 1} = \underset{\sim}{\mathfrak{K}}_1(\underset{\sim}{H}_1), \tag{15.7}$$

where $\underset{\sim}{\mathfrak{K}}_1$ is a linear function.

Exercise 15.1. Prove that

$$L[\underset{\sim}{\tilde{E}}] = \underset{\sim}{\mathfrak{K}}_1(\underset{\sim}{H}_1), \tag{15.8}$$

where L is the linear elasticity (cf. (14.14)).

The precise form of $\underset{\sim}{T}_{\underset{\sim}{\kappa} n}$ is not needed now. It is essential to observe only that if we set

$$\underset{\sim}{\tilde{E}}_n \equiv \frac{1}{2} (\underset{\sim}{H}_n + \underset{\sim}{H}_n^T), \tag{15.9}$$

then by (15.8)

$$\underset{\sim}{T}_{\kappa n} = L[\tilde{\underset{\sim}{E}}_n] + \underset{\sim}{\mathfrak{h}}_n(\underset{\sim}{H}_1, \underset{\sim}{H}_2, \ldots, \underset{\sim}{H}_{n-1}), \tag{15.10}$$

where $\underset{\sim}{\mathfrak{h}}_n$ is a function we need not know explicitly in what follows. That is, the infinitesimal strain $\tilde{\underset{\sim}{E}}_n$ corresponding to the nth term in the expansion of $\underset{\sim}{H}$ enters by itself and in just the same way as $\tilde{\underset{\sim}{E}}$ enters the stress relation of the infinitesimal theory.

We shall use Cauchy's first law of motion in the form $(11.14)_1$:

$$\text{Div}\ \underset{\sim}{T}_{\kappa} + \rho_{\kappa}\underset{\sim}{b} = \rho_{\kappa}\ddot{\underset{\sim}{x}} = \rho_{\kappa}\ddot{\underset{\sim}{u}}. \tag{11.14$_1$}$$

Substituting (15.6), $(15.3)_1$, (15.5), and (15.10) into this equation and equating like powers of ε, and doing likewise with the traction boundary condition $(15.3)_2$, we find the following <u>iterative system</u> for determining the $\underset{\sim}{u}_n$ successively:

$$\begin{aligned} \text{Div}\ L[\tilde{\underset{\sim}{E}}_n] + \rho_{\kappa}\underset{\sim}{b}^*_n &= \rho_{\kappa}\ddot{\underset{\sim}{u}}_n \quad \text{in } \mathcal{B}_{\kappa}, \\ L[\tilde{\underset{\sim}{E}}_n]\underset{\sim}{n}_{\kappa} &= \underset{\sim}{t}^*_{\kappa n} \quad \text{on } \partial\mathcal{B}_{\kappa}, \end{aligned} \tag{15.11}$$

where

$$\begin{aligned} \rho_{\kappa}\underset{\sim}{b}^*_n &= \rho_{\kappa}\underset{\sim}{b}_n + \text{Div}\ \underset{\sim}{\mathfrak{h}}_n(\underset{\sim}{H}_1, \underset{\sim}{H}_2, \ldots, \underset{\sim}{H}_{n-1}), \\ \underset{\sim}{t}^*_{\kappa n} &= \underset{\sim}{t}_{\kappa n} - \underset{\sim}{\mathfrak{h}}_n(\underset{\sim}{H}_1, \underset{\sim}{H}_2, \ldots, \underset{\sim}{H}_{n-1})\underset{\sim}{n}_{\kappa}, \\ n &= 1,2,3, \ldots \end{aligned} \tag{15.12}$$

The system (15.11) has exactly the form that defines the traction boundary-value problem in the infinitesimal theory. Thus, it seems, the solution of the traction boundary-value problem in the finite theory, granted the assumptions (15.2), (15.3), and (15.4) to the nth order in ε, is reduced to the solution of n traction boundary-value problems in the infinitesimal theory, the first of which is the corresponding classical boundary-value problem, for the same body. The loads for the nth-order problem are $\underset{\sim}{b}^*_n$

and $\underset{\sim}{t}_n^*$. By (15.12), these are not merely the nth terms in the given loads (15.3); rather, they are certain functions of the solutions $\underset{\sim}{u}_1$, $\underset{\sim}{u}_2$, $\underset{\sim}{u}_3$, ..., $\underset{\sim}{u}_{n-1}$ of the preceding $n-1$ stages of the process.

In general, perturbation methods yield solutions only for problems not essentially different from linear ones. Elasticity is no exception here. Consider, for example, the static problem when $\underset{\sim}{b} = \underset{\sim}{0}$ and $\underset{\sim}{t}_{\underset{\sim}{\kappa}} = \underset{\sim}{0}$. The system (15.11), (15.12) then reduces for every n to the corresponding problem of the infinitesimal theory. By the uniqueness theorem of that theory, $\underset{\sim}{u}_n = \underset{\sim}{0}$ for every n. Hence the only solution is the trivial one. However, the finite theory should be able to handle the problem of eversion, in which at least two solutions exist when $\underset{\sim}{b} = \underset{\sim}{0}$ and $\underset{\sim}{t}_{\underset{\sim}{\kappa}} = \underset{\sim}{0}$, and the theory ought to determine the non-trivial one, which gives the everted shape. We shall return, however, to considering such results as the perturbation method yields.

When we arrive at (15.11) and (15.12), we face a difficulty in the equilibrium problem, $\ddot{\underset{\sim}{u}} = \underset{\sim}{0}$. In the infinitesimal theory, a <u>necessary condition</u> for solution of the traction boundary-value problem is that the applied loads $\underset{\sim}{b}$ and $\underset{\sim}{t}_{\underset{\sim}{\kappa}}$ form an equilibrated system. That is, the resultant force and resultant moment of force corresponding to these loads must vanish. Otherwise, no solution exists. In the infinitesimal theory, this point occasions no trouble, since it compels us merely not to set up any traction boundary-value problems when the applied loads are not equilibrated. Similarly, in the general theory, the loads must be equilibrated if a solution to the traction boundary-value problem is to exist. Thus we assume that $\underset{\sim}{b}$ and $\underset{\sim}{t}_{\underset{\sim}{\kappa}}$ are equilibrated in $\mathcal{B}_{\underset{\sim}{\kappa}}$, and consequently, by (15.3), $\varepsilon = \underset{\sim}{b}_n$ and $\underset{\sim}{t}_{\underset{\sim}{\kappa}n}$ are equilibrated in $\mathcal{B}_{\underset{\sim}{\kappa}}$, for each n. But these are not the loads for the iterative problem defined by (15.11) and (15.12). Rather, the loads are $\underset{\sim}{b}_n^*$ and $\underset{\sim}{t}_{\underset{\sim}{\kappa}n}^*$, which depend upon $\underset{\sim}{u}_1$, $\underset{\sim}{u}_2$, ..., $\underset{\sim}{u}_{n-1}$. Thus in order that (15.11) and (15.12) have a solution $\underset{\sim}{u}_n$ at all, it is necessary that a condition be satisfied by the solution $\underset{\sim}{u}_{n-1}$ found at the previous stage.

<u>Exercise 15.2</u>. Show that the resultant force exerted by the loads (15.12) vanishes.

Consequently, the only condition of compatibility for the nth-order problem is that the resultant moment of the loads $\mathbf{b}^*_n$ and $\mathbf{t}^*_{\kappa n}$ shall vanish. In order that the problem of the exact theory have a solution at all, it is necessary that for any fixed point $\mathbf{x}_0$ in $\mathcal{B}_{\chi}$

$$\int_{\partial\mathcal{B}_{\chi}} (\mathbf{x} - \mathbf{x}_0)\times\mathbf{t}\,ds + \int_{\mathcal{B}_{\chi}} (\mathbf{x} - \mathbf{x}_0)\times\mathbf{b}\,dm = \mathbf{0}. \tag{15.13}$$

Equivalently,

$$\int_{\partial\mathcal{B}_{\kappa}} (\mathbf{x} - \mathbf{x}_0)\times\mathbf{t}_{\kappa}\,ds + \int_{\mathcal{B}_{\kappa}} (\mathbf{x} - \mathbf{x}_0)\times\mathbf{b}\,dm = \mathbf{0}. \tag{15.14}$$

Since $\mathbf{x} = \mathbf{X} + \mathbf{u}$ and $\mathbf{u}$ is unknown, this condition cannot be used at once as a criterion for the assigned loads $\mathbf{t}_{\kappa}$ and $\mathbf{b}$.

Exercise 15.3. Prove da SILVA's theorem: If the resultant force exerted by the loads $\mathbf{t}_{\kappa}$ and $\mathbf{b}$ vanishes, then there exists a rotation carrying κ into a configuration κ^* in which the resultant moment of the load with respect to any point $\mathbf{X}_0$ vanishes:

$$\int_{\partial\mathcal{B}_{\kappa^*}} (\mathbf{X} - \mathbf{X}_0)\times\mathbf{t}_{\kappa}\,ds + \int_{\mathcal{B}_{\kappa^*}} (\mathbf{X} - \mathbf{X}_0)\times\mathbf{b}\,dm = \mathbf{0}. \tag{15.15}$$

Since a rotation of the reference configuration preserves any important property that configuration might have, such as being natural, undistorted, or homogeneous, we lose no essential generality if we assume κ to be so chosen that (15.15) holds. That is, we assume that the given loads are such as to exert neither resultant force nor resultant moment on $\mathcal{B}_{\kappa}$. Of course, this is not at all the same thing as (15.14), which contains the unknown place $\mathbf{x}$ rather than the given place $\mathbf{X}$.

In effect dropping the $\mathbf{X}_o$ from (15.15), we subtract that equation from (15.14) and obtain

$$\int_{\partial\mathcal{B}_{\kappa}} \mathbf{u}\times\mathbf{t}_{\kappa}\,ds + \int_{\mathcal{B}_{\kappa}} \mathbf{u}\times\mathbf{b}\,dm = \mathbf{0}. \tag{15.16}$$

Substituting the series (15.13) and (15.14) into (15.16) and equating to zero the coefficient of each power of ε, we obtain SIGNORINI's conditions of compatibility:

$$\sum_{q=1}^{m-1} \left[\int_{\partial\mathcal{B}_{\kappa}} \mathbf{u}_{m-q} \times \mathbf{t}_{\kappa q}\, ds + \int_{\mathcal{B}_{\kappa}} \mathbf{u}_{m-q} \times \mathbf{b}_{q}\, dm\right] = 0, \qquad m = 2, 3, \ldots \tag{15.17}$$

That is, in order that the equations (15.11) be soluble for the displacement $\mathbf{u}_n$, it is necessary that the displacements $\mathbf{u}_1, \mathbf{u}_2, \ldots, \mathbf{u}_{n-1}$ satisfy (15.17) for $m = 2, 3, \ldots, n$.

The conditions (15.17) have been derived as a necessary and sufficient condition that the loads acting on the body in χ shall be equilibrated. We expect that they should likewise be the formal conditions that the successive infinitesimal problems (15.11) be compatible. This is so.

Exercise 15.4. Prove that the effective loads $\mathbf{t}^*_n$, $\mathbf{b}^*_n$ as given by (15.12) are equilibrated if and only if (15.17) holds.

In the infinitesimal theory, the solution of a traction boundary-value problem, if it exists, is unique only to within an infinitesimal rotation. Thus if $\mathbf{u}_n(\mathbf{X})$ is a solution of the system (15.11), so also is $\mathbf{u}_n + \boldsymbol{\omega}_n (\mathbf{X} - \mathbf{X}_0) + \text{const.}$, where $\boldsymbol{\omega}_n$ is any constant vector. In the general theory of elasticity, no such indeterminacy is to be expected, for an arbitrary rotation of the body as a whole does not generally preserve the equilibrium of moments. To reconcile these facts, SIGNORINI proposed to determine $\boldsymbol{\omega}_n$ by the compatibility condition (15.17). If a unique $\boldsymbol{\omega}_n$ exists, the indeterminacy of the rotation is removed.

Let us suppose, then, that solutions $\mathbf{u}_1, \mathbf{u}_2, \ldots, \mathbf{u}_{p-1}$ of (15.11) have been found when $n = 1, 2, \ldots, p-1$ such that the conditions of compatibility (15.17) are satisfied if $m = 2, 3, \ldots, p$. The system (15.11) is then compatible when $n = p$. Let $\mathbf{u}_p$ be any particular solution. Put

$$\mathbf{r}_p \equiv \sum_{q=1}^{p} \int_{\partial\mathcal{B}_{\kappa}} \mathbf{u}_{p+1-q} \times \mathbf{t}_{\kappa q}\, ds + \int_{\mathcal{B}_{\kappa}} \mathbf{u}_{p+1-q} \times \mathbf{b}_{q}\, dm]. \tag{15.18}$$

The vector $\underset{\sim}{r}_p$ is defined in terms of the assigned loads (15.3) and the functions $\underset{\sim}{u}_1, \underset{\sim}{u}_2, \ldots, \underset{\sim}{u}_p$. Since $\underset{\sim}{u}_p$ is a particular solution of (15.11), every other solution is of the form $\underset{\sim}{u}_p + \underset{\sim}{\omega}_p \times (\underset{\sim}{X} - \underset{\sim}{X}_0) +$ const., where $\underset{\sim}{\omega}_p$ is a constant vector. The problem is now to determine an $\underset{\sim}{\omega}_p$ such that the compatibility condition (15.17) is satisfied when $m = p + 1$. In terms of the notation (15.18), this condition may be written as

$$\int_{\partial \mathcal{B}_{\underset{\sim}{\kappa}}} (\underset{\sim}{\omega}_p \times (\underset{\sim}{X} - \underset{\sim}{X}_0)) \times \underset{\sim}{t}_{\underset{\sim}{\kappa}1}\, ds + \int_{\mathcal{B}_{\underset{\sim}{\kappa}}} (\underset{\sim}{\omega}_p \times (\underset{\sim}{X} - \underset{\sim}{X}_0)) \times \underset{\sim}{b}_1\, dm = -\underset{\sim}{r}_p . \tag{15.19}$$

Since $(\underset{\sim}{a} \times \underset{\sim}{b}) \times \underset{\sim}{c} = (\underset{\sim}{a} \cdot \underset{\sim}{c})\underset{\sim}{b} - (\underset{\sim}{b} \cdot \underset{\sim}{c})\underset{\sim}{a}$, this condition may be put into the form

$$(\underset{\sim}{A}_{(1)} - \underset{\sim}{1}\, \mathrm{tr}\, \underset{\sim}{A}_{(1)})\underset{\sim}{\omega}_p = -\underset{\sim}{r}_p, \tag{15.20}$$

where

$$\underset{\sim}{A}_{(1)} \equiv \int_{\partial \mathcal{B}_{\underset{\sim}{\kappa}}} (\underset{\sim}{X} - \underset{\sim}{X}_0) \otimes \underset{\sim}{t}_{\underset{\sim}{\kappa}1}\, ds + \int_{\mathcal{B}_{\underset{\sim}{\kappa}}} (\underset{\sim}{X} - \underset{\sim}{X}_0) \otimes \underset{\sim}{b}_1\, dm. \tag{15.21}$$

Exercise 15.5. $\underset{\sim}{A}_{(1)}$ is called the astatic load corresponding to the loads $\underset{\sim}{t}_{\underset{\sim}{\kappa}1}$, $\underset{\sim}{b}_1$. A line such that rotation of $\mathcal{B}_{\underset{\sim}{\kappa}}$ through any angle about it, while $\underset{\sim}{t}_{\underset{\sim}{\kappa}1}$ and $\underset{\sim}{b}_1$ are kept fixed, yields an equilibrated system of forces on $\mathcal{B}_{\underset{\sim}{\kappa}}$, is said to be an axis of equilibrium for the loads on $\mathcal{B}_{\underset{\sim}{\kappa}}$. Prove that an axis of equilibrium exists for the loads on $\mathcal{B}_{\underset{\sim}{\kappa}}$ if and only if

$$\det\, (\underset{\sim}{A}_{(1)} - \underset{\sim}{1}\, \mathrm{tr}\, \underset{\sim}{A}_{(1)}) = 0. \tag{15.22}$$

The linear equation (15.20) has a unique solution $\underset{\sim}{\omega}_p$ if and only if

$$\det\, (\underset{\sim}{A}_{(1)} - \underset{\sim}{1}\, \mathrm{tr}\, \underset{\sim}{A}_{(1)}) \neq 0. \tag{15.23}$$

Therefore, the analysis so far proves SIGNORINI's theorem of compatibility and uniqueness: Let it be supposed that the loads $\underset{\sim}{t}_{\underset{\sim}{\kappa}1}$, $\underset{\sim}{b}_1$

<u>do not</u> possess an axis of equilibrium. Then

1. If solutions exist for the general traction boundary-value problem in the infinitesimal theory, solutions $\underset{\sim}{u}_1, \underset{\sim}{u}_2, \ldots$ exist for the iterative system (15.11).

2. If the classical uniqueness theorem for the traction boundary-value problem of the infinitesimal theory holds, then the solutions $\underset{\sim}{u}_1, \underset{\sim}{u}_2, \ldots$ are unique to within uniform translations.

If the loads do possess an axis of equilibrium, (15.23) fails, and (15.20) may have more than one solution $\underset{\sim}{\omega}_p$, or no solution. The status of the method is then a delicate matter. It is clear that if an axis of equilibrium exists, the load cannot be expected to determine a <u>unique</u> solution, since rotations about the axis may well carry the body from one configuration into another subject to exactly the same forces and hence indistinguishable from it.

SIGNORINI's method has been reformulated by RIVLIN & TOPAKOGLU in terms perhaps easier to interpret. They suggested that since the step from any one stage to the next is a small one, for small enough ε, the discrepancy between the successive solutions is equivalent to the effect of a set of forces, which are cancelled by supplying the displacement that corresponds, according to the infinitesimal theory, to their negatives. This idea they rendered precise. By use of it, some special solutions have been found. The most important of these concern the second-order theory of isotropic materials.

<u>Exercise 15.6</u>. Prove that as far as second-order terms in $\underset{\sim}{H}$, the stress relation for isotropic materials assumes the form

$$\frac{\underset{\sim}{T}}{\mu} = \frac{\lambda}{\mu} I\underset{\sim}{1} + 2\underset{\sim}{\tilde{E}} + \left(\frac{1}{2}\frac{\lambda}{\mu}\operatorname{tr}(\underset{\sim}{H}\underset{\sim}{H}^T) + \alpha_3 I^2 + \alpha_4 II\right)\underset{\sim}{1} + \alpha_5 I\underset{\sim}{\tilde{E}} + \underset{\sim}{H}\underset{\sim}{H}^T + \alpha_6\underset{\sim}{\tilde{E}}^2, \tag{15.24}$$

where $\underset{\sim}{\tilde{E}} \equiv \frac{1}{2}(\underset{\sim}{H} + \underset{\sim}{H}^T)$, $I \equiv \operatorname{tr}\underset{\sim}{\tilde{E}}$, $2II \equiv (\operatorname{tr}\underset{\sim}{\tilde{E}})^2 - \operatorname{tr}\underset{\sim}{\tilde{E}}^2$, and α_3, α_4, α_5, and α_6 are dimensionless constants called <u>second-order elasticities</u>.

By using the above-stated method, RIVLIN calculated the general solution of POYNTING's problem for a cylinder of arbitrary cross-section.

Let the cross-section of the cylinder before deformation have area, classical torsional rigidity, and polar moment of inertia A_o, S_o, and I_o, respectively. Let ε be the twist. Then RIVLIN found that the volume average of the extension $v - 1$ produced by a twist of amount ε is given be

$$\bar{v} - 1 = \frac{\varepsilon^2}{2A_o}\left[-\frac{2\beta''}{1+\sigma} S_o - (I_o - S_o)\right], \tag{15.25}$$

where

$$\sigma \equiv \frac{\lambda}{2(\lambda + \mu)},$$

$$\beta'' = \frac{1}{8}\left[-(1 - 2\sigma)\alpha_4 + (1 - \sigma)\alpha_6\right].$$

No general rule of sign can be inferred from (15.25). Whether torsion effects lengthening or shortening depends on the value of the dimensionless modulus β''.

Likewise, the magnitude of the KELVIN effect is determined from the second-order elasticities. For a circular-cylindrical tube, the fractional change of volume effected by a twist of amount ε may be shown to be

$$\frac{\Delta v}{v} = \pi\varepsilon^2\beta'''(R_0^2 + R_1^2), \tag{15.27}$$

where R_0 and R_1 are the initial inner and outer radii, and where

$$\beta''' \equiv -\frac{1}{4} + \frac{1}{16(1+\sigma)}\left[(3 - 4\sigma - 4\sigma^2)\alpha_4 + (2\sigma^2 + 2\sigma - 2)\alpha_6\right]. \tag{15.28}$$

To derive these results is fairly straightforward though lengthy. The method of calculation is not of particular interest, but the results are of great importance. They fully explain the main second-order effects of elasticity, and they show that the magnitudes of those effects cannot be determined from the classical moduli λ and μ or from each other. The dimensionless constants β'' and β''' are determined by λ/μ and two of the second-order elasticities. The results (15.25) - (15.28) show that experiments measuring the second-order elongation and change of bulk suffice to determine α_4 and α_6 in a material for which σ is known.

LECTURE 16: INEQUALITIES FOR ISOTROPIC MATERIALS.

Inequalities in the Infinitesimal Theory.

As mentioned several times already, it is customary to impose a priori restrictions on the linear elasticity L in the stress relation (14.14) of the classical infinitesimal theory. These are, first, that

$$L = L^T , \tag{16.1}$$

reducing the maximum number of independent components of L from 36 to 21, and that

$$\mathrm{tr}(L[\tilde{E}]\tilde{E}) > 0 \quad \text{if} \quad \tilde{E} \neq 0 . \tag{16.2}$$

For isotropic materials by (14.15), the condition (16.1) is satisfied automatically, while (16.2) is equivalent to

$$\mu > 0 , \quad 3\lambda + 2\mu > 0 . \tag{16.3}$$

If (16.1) holds, the condition (16.2) asserts that in any non-rigid infinitesimal deformation from the natural state, positive work is done.

The conditions (16.1) and (16.2) are sufficient that the mixed boundary-value problem of place and traction have a solution for sufficiently smooth domains and data, and that that solution be stable and be unique to within an infinitesimal rotation. If these conditions are relaxed, a solution will fail to exist for at least some domains and some data.

The inequality (16.2) may be motivated in five ways, the first three of which have been mentioned already:

1. It is a necessary and sufficient condition, in general, for existence and uniqueness of solution to the standard boundary-value problems.

2. It is a sufficient condition for stability of the solution.

3. It represents, if somewhat vaguely, an assumption made plausible by thermodynamics.

4. It is a sufficient condition that the speeds corresponding to all types of waves be real and non-vanishing.

5. It is equivalent to plausible requirements on classes of static strains. Indeed, for an isotropic material (14.15) yields

$$T\langle 12\rangle = 2\mu\tilde{E}\langle 12\rangle \quad , \qquad \operatorname{tr}\underset{\sim}{T} = (3\lambda + 2\mu)\operatorname{tr}\underset{\sim}{\tilde{E}} \ , \tag{16.4}$$

where the physical components in the first equation are taken with respect to any pair of orthogonal directions.

Exercise 16.1. Letting K be the amount of shear, in the planes X_3 = const., of the planes X_1 = const. in the X_2-direction at $\underset{\sim}{X}$ (cf. (5.11)), prove that when K is small

$$2\tilde{E}\langle 12\rangle = K \ . \tag{16.5}$$

Letting $\Delta\upsilon$ be the increment of specific volume υ at $\underset{\sim}{X}$ produced by an infinitesimal deformation, prove that

$$\operatorname{tr}\underset{\sim}{\tilde{E}} = \frac{\Delta\upsilon}{\upsilon} \ . \tag{16.6}$$

Hence interpret $(16.4)_1$ as a statement that the shear stress in any infinitesimal deformation shall point in the direction of the shear effected, and $(16.4)_2$ as a statement that pressure in mean is required to decrease the volume, tension in mean to increase it.

By (16.4), in order to conclude that (16.3) should hold, it is not necessary to consider all infinitesimal deformations. It suffices to require that in some one simple shear with positive K, the corresponding shear stress $T\langle 12\rangle$ shall be positive, and that in some one dilatation $\underset{\sim}{u} = \alpha\underset{\sim}{X}$ with positive α, the corresponding stress shall be a uniform, non-vanishing tension.

The Problem for Finite Deformations.

The foregoing considerations show that in order to get plausible results from any theory of elasticity, some sort of *a priori* inequality is necessary. It is not sufficient just to require that every response function

$\mathfrak{g}_K$ in the general theory defined by (11.2) shall be such as to yield (16.2) in infinitesimal deformations. In such a material it would still be possible that a bar if pulled by a great enough tensile force would shorten, or that a sphere if subjected to great hydrostatic pressure should expand. Some a priori inequality beyond (16.2) is clearly necessary, but what this inequality should be remains a question for study and debate.

Arguments 1 and 2 do not apply in any straightforward way. As we have seen in Lecture 11, unqualified uniqueness of solutions to the mixed boundary-value problem is not wished, so no inequality strong enough to deliver it is a desirable *a priori* inequality. One of the principal aims of the theory of finite elastic strain is to yield criteria of instability, so no inequality strong enough to insure that all solutions are stable ones is fit to be laid down as general. Argument 3 cannot be made precise without the introduction of a true, open thermodynamics of deformation. In a later lecture we shall present such a thermodynamics and its consequences, but in adopting them the student must take on the stronger burden of accepting a new theory far more daring than elasticity in order to get inequalities of a purely mechanical nature. Argument 4, also, we shall consider later in the context of the finite theory, but even in the infinitesimal theory it is not adequate, since (16.2) is merely sufficient, not necessary, for real wave speeds. *E.g.*, for isotropic materials necessary and sufficient conditions are the weaker inequalities $\mu > 0$, $\lambda + 2\mu > 0$. Thus Argument 4 can never be developed into an adequate basis for a general *a priori* inequality.

In this lecture and the next we shall consider forms that Argument 5 may take for the finite theory.

Some Simple Static Inequalities.

We shall consider an isotropic elastic material in an undistorted state and take the stress relation in the form (11.38), giving the principal stresses t_i as functions of the principal stretches v_k:

$$t_i = t_i(v_1, v_2, v_3), \qquad i = 1, 2, 3 . \tag{11.38}$$

Alternatively, we may use the principal forces T_i as given by (11.41) and obtain

$$T_i = T_i(v_1, v_2, v_3) \quad , \qquad i = 1, 2, 3, \tag{16.7}$$

where of course either set of three functions determines the other.

If we hold two principal stretches fixed but increase the third, it is natural to expect that tension must be applied; if instead we decrease the third, we expect to have to apply pressure. This expectation may be expressed as the <u>tension-extension inequality</u> (T-E):

$$(\overset{*}{T}_i - T_i)(\overset{*}{v}_i - v_i) > 0 \quad \text{or} \quad (\overset{*}{t}_i - t_i)(\overset{*}{v}_i - v_i) > 0 \tag{16.8}$$

$$\text{if} \quad \overset{*}{v}_j \neq v_j \qquad \text{when} \qquad j \neq i \quad ,$$

where $\overset{*}{T}_i \equiv T_i(\overset{*}{v}_1, \overset{*}{v}_2, \overset{*}{v}_2)$, $\overset{*}{t}_i \equiv t_i(\overset{*}{v}_1, \overset{*}{v}_2, \overset{*}{v}_3)$. Either statement of this inequality is equivalent to the other.

It is natural to ask if the relations (11.38) and (16.7) be invertible. Clearly, in general, (11.38) cannot be, for in the special case of an elastic fluid they never are: Since $t_1 = t_2 = t_3$ always in a fluid, unique values of the v_1, v_2, v_3 cannot correspond to assigned values of the t_i, and in fact only the product $v_1 v_2 v_3$ is determined by the t_i. For (16.7), however, it is a different matter, and there is nothing in the experiential background that suggests an objection against requiring the principal stretches to be determined uniquely by the principal forces T_i.

<u>Exercise 16.2</u>. A <u>perfect gas</u> is defined by the constitutive equation

$$\underset{\sim}{T} = -p\underset{\sim}{1} \quad , \qquad p = K v_1 v_2 v_3, \quad K > 0 \ . \tag{16.9}$$

Prove that in a perfect gas the relations (16.7) are invertible at every argument v_1, v_2, v_3.

In general, if the relations (16.7) are invertible, so that

$$v_i = v_i(T_1, T_2, T_3) \quad , \tag{16.10}$$

we shall say that the <u>IFS-condition</u> is satisfied, where "IFS" is a mnemonic for "invertibility of force-stretch".

Granted (16.10), it is natural to put $v_i^* \equiv v_i(T_1^*, T_2^*, T_3^*)$ and to demand that increasing one principal force T_i while holding the others constant shall produce an increase in the corresponding stretch:

$$(T_i^* - T_i)(v_i^* - v_i) > 0 \quad \text{if} \quad T_j^* = T_j \quad \text{when} \quad j \neq i \,. \tag{16.11}$$

This condition may be called the <u>extension-tension inequality</u> (E-T).

The E-T and T-E inequalities refer to variation of a single stretch or a single force. When we compare two principal stretches or when we compare two principal forces, it is natural to expect that the greater one correspond to the greater stretch. The <u>Baker-Ericksen inequality</u> (B-E) asserts that

$$(t_i - t_j)(v_i - v_j) > 0 \quad \text{if} \quad v_i \neq v_j \,, \tag{16.12}$$

while the <u>ordered forces inequality</u> (O-F) asserts that

$$(T_i - T_j)(v_i - v_j) > 0 \quad \text{if} \quad v_i \neq v_j \,. \tag{16.13}$$

The two are not equivalent, and there is no clear reason for preferring one to the other. Clearly the B-E inequality cannot be satisfied in an elastic fluid, since $t_a = t_b$, independently of whatever the v_i may be. The O-F inequality can be satisfied by a fluid, as shown in the following exercise.

<u>Exercise 16.3</u> Prove that in a perfect gas the O-F inequality is satisfied for every argument v_1, v_2, v_3 .

Connections between the B-E and O-F inequalities can be found. In fact, it can be shown that for any values of the t_i, one or other but not both of the implications O-F $\Rightarrow$ B-E and B-E $\Rightarrow$ O-F hold.

<u>Exercise 16.4</u>. Let $\beth_r$ be the response coefficients in the representation (11.35). The E-<u>inequalities</u> are

$$\beth_0 \leqq 0 \,, \quad \beth_1 > 0 \,, \quad \beth_{-1} \leqq 0 \,. \tag{16.14}$$

(The E-inequalities are never satisfied by an elastic fluid, since for it $\beth_1 = 0$.) Prove that

$$\text{E} \Rightarrow \text{B-E} \;\&\; \text{O-F} \,. \tag{16.15}$$

Finally, suppose all three stretches are equal: $v_1 = v_2 = v_3 = v$. By (11.38), $t_1 = t_2 = t_3 = t$, say, where $t = t(v)$, and also $T_1 = T_2 = T_3 = T$, say, where $T = T(v) = v^2 t(v)$. If the undistorted state is a natural state, so that $t(1) = 0$, it is natural to expect that tension is required to effect increase in volume, compression to effect decrease, so that

$$t(v - 1) > 0 \ , \qquad T(v - 1) > 0 \ . \tag{16.16}$$

The two conditions are equivalent. If the undistorted state is not a natural state, however, the more general conditions

$$(t^* - t)(v^* - v) > 0 \ , \qquad (T^* - T)(v^* - v) > 0 \ , \tag{16.17}$$

are not equivalent. We shall refer to the latter as the <u>pressure-compression</u> inequality (P-C).

<u>Exercise 16.5.</u> Prove that in a state of tension, $(16.17)_1$ implies $(16.17)_2$, while in a state of pressure, the reverse implication follows. Prove that in an elastic fluid, $(16.17)_1$ reduces to an assertion that the pressure p is an increasing function of $\rho^{1/3}$ and hence of ρ, while $(16.17)_2$ reduces to an assertion that $\rho^{-2/3}p(\rho)$ is an increasing function of ρ, <u>i.e.</u>

$$\rho \frac{dp}{d\rho} \geqq \frac{2}{3} p \ . \tag{16.18}$$

This last result shows that the P-C inequality, when referred to fluids, imposes a strong and not always plausible restriction upon the pressure $p(\rho)$. The inequality (16.18) requires that the compressibility of a gas cannot vanish unless the pressure does. Such is the case for a perfect gas (Exercises 16.2 and 16.3) and more generally for a "polytropic" fluid, namely, a fluid having the constitutive equation $p = K\rho^\gamma$, $K > 0$, $\gamma \geqq 1$, but not for a fluid capable of change of phase, <u>e.g.</u> a van der Waals fluid.

<u>Exercise 16.6.</u> Prove that the foregoing conditions assume the following forms in the infinitesimal theory:

T-E: $\lambda + 2\mu > 0$

IFS: $\mu(3\lambda + 2\mu) \neq 0$

E-T: $\dfrac{\mu(3\lambda + 2\mu)}{\lambda + \mu} > 0$

$$\text{O-F}, \text{B-E}, \text{E}: \mu > 0$$

$$\text{P-C}: \mu > 0 , \quad (3\lambda + 2\mu) > 0 .$$

The GCN_0 Condition.

All the foregoing conditions except IFS assert that some function is monotone increasing. They suggest, more generally, that the transformation (16.7) from principal stretches v_i to principal forces T_i may be a monotone transformation:

$$\sum_{i=1}^{3} (T_i^* - T_i)(v_i^* - v_i) > 0 \quad \text{unless} \quad v_k^* = v_k, \ k = 1,2,3 . \qquad (16.19)$$

This inequality is called the GCN_0 condition. We shall verify now that it includes as special cases a number of the foregoing.

1. Suppose $v_b^* = v_b$ except when $b = a$. Then the T-E inequality (16.8) follows.

2. If $v_i^* \neq v_i$ for any i, it is impossible that $T_i^* = T_i$ for all i. Thus the IFS condition (16.10) follows.

3. In view of the IFS condition, we may substitute (16.10) into (16.19), whence (16.11) follows.

4. Let $\pi(i)$ denote a permutation of the numbers 1, 2, 3. By (11.39) and (11.41), a permutation of the v_i induces a permutation of the T_i. That is, if $v_i^* = v_{\pi(i)}$, then $T_i^* = T_{\pi(i)}$. Hence (16.19) yields as a special case

$$\sum_{i=1}^{3} (T_{\pi(i)} - T_i)(v_{\pi(i)} - v_i) > 0 \quad , \qquad (16.20)$$

If we choose $\pi(i)$ as a permutation of j and k with ℓ held fixed, (16.20) reduces to (16.13).

5. If $v_1 = v_2 = v_3$, by (11.30) and (11.41) it follows that $T_1 = T_2 = T_3$, so that (16.19) reduces to (16.17).

What we have proved, then, may be stated schematically as follows:

$$GCN_0 \Rightarrow \begin{cases} \text{T-E} & \& \\ \text{IFS} & \& \\ \text{E-T} & \& \\ \text{O-F} & \& \\ \text{P-C} & \end{cases} \qquad (16.21)$$

That is, the GCN_0 condition is sufficient to ensure all the inequalities mentioned above as being plausible, except for the B-E inequality. In general, it is not possible to reverse the above implication.

<u>Exercise 16.7</u>. Set

$$J_{ab} \equiv \partial_{v_b} T_a \quad . \tag{16.22}$$

Prove that GCN_0 implies that the quadratic form of J_{ab} be non-negative definite. The stronger condition that it be positive -definite is called the <u>GCN_0^+ condition</u>. Clearly $GCN_0^+ \Rightarrow GCN_0$. Prove that

$$GCN_0^+ \Rightarrow \begin{cases} T\text{-}E^+ & \& \\ IFS^+ & \& \\ E\text{-}T^+ & \& \\ O\text{-}F & \& \\ P\text{-}C & \end{cases} \tag{16.23}$$

where $T\text{-}E^+$, IFS^+, and $E\text{-}T^+$ are conditions stronger than T-E, IFS, and E-T, respectively:

$$\partial_{v_i} T_i > 0 \ , \ \det \|J_{ab}\| \neq 0 \ , \ \partial_{T_i} v_i > 0 \ . \tag{16.24}$$

If, further,

$$J_{ab} = J_{ba} \tag{16.25}$$

prove that

$$\left.\begin{matrix} C & \& \\ IFS^+ & \& \\ T\text{-}E^+ & \& \\ E\text{-}T & \end{matrix}\right\} \Leftrightarrow GCN_0^+ \tag{16.26}$$

where C stands for (16.3). Prove also that the four inequalities symbolized on the left-hand side are independent.

This last result is particularly telling. In the infinitesimal theory, by the result of Exercise (16.6), slightly strengthened,

$$C \Rightarrow T\text{-}E^+ \ \& \ IFS^+ \ \& \ E\text{-}T^+ \ \& \ O\text{-}F \ \& \ B\text{-}E \ \& \ E \ \& \ P\text{-}C \ . \tag{16.27}$$

In fact, C $\Leftrightarrow$ P-C. Even in the special case when (16.25) holds, four of the inequalities listed on the right-hand side of (11.26) are independent. If they all hold, we may express this fact by the convenient and simple GCN_0^+ condition, which in turn suffices to imply the two further inequalities O-F and C.

<u>Exercise 16.8</u>. Using (11.44), show that

$$\text{T-E}^+ \Leftrightarrow F<1111> > 0 \ \& \ F<2222> > 0 \ \& \ F<3333> > 0 \ , \qquad (16.28)$$

while

$$\text{B-E} \Leftrightarrow F<1212> > 0 \ , \quad F<2323> > 0 \ , \quad F<3131> > 0 \ . \qquad (16.29)$$

LECTURE 17: A GENERAL INEQUALITY.

Problem

In the last lecture we remarked that some a priori inequality is required in elasticity in order to get the results of the classical infinitesimal theory, generally accepted as just for small deformations. On the other hand, we cannot follow blindly the lead of pure analysis and impose conditions strong enough to yield unqualified uniqueness of solution to the boundary-value problem of place and traction, since such uniqueness in large strain would be just as inappropriate as failure of that uniqueness in small. In any case, this caution is presently an empty one, since the general differential equations of elasticity lie outside the domain for which analysts have constructed a theory.

As a summary of plausible static requirements for an isotropic elastic material in an undistorted state we set up for study the GCN_0 condition:

$$\sum_{i=1}^{3} (T_i^* - T_i)(v_i^* - v_i) > 0 \qquad (16.18)$$

unless $v_i^* = v_i$ when $i = 1, 2, 3$. This condition asserts that the transformation from principal stretches to principal forces is monotone.

Our problem now is to find a corresponding condition for general elastic materials.

The GCN Inequality.

A possible generalization of (16.18) is the requirement that the transformation from the deformation gradient $\underset{\sim}{F}$ to the Piola stress $\underset{\sim}{T}_{\kappa}$ be monotone:

$$\text{tr}\,\{(\underset{\sim}{T}_{\kappa}^* - \underset{\sim}{T}_{\kappa})(\underset{\sim}{F}^{*T} - \underset{\sim}{F}^{T})\} > 0 \qquad (17.1)$$

unless $\underset{\sim}{F}^* = \underset{\sim}{F}$. Here we are using the stress relation in the form (11.15):

$$\mathbf{T}_{\kappa} = \mathfrak{h}(\mathbf{F}) \ , \tag{11.15}$$

and $\mathbf{T}^*_{\kappa} \equiv \mathfrak{h}(\mathbf{F}^*)$. The condition (17.1), however, is not frame-indifferent.

<u>Exercise 17.1</u>. Suppose $\mathbf{F}^* = \mathbf{QF}$, where $\mathbf{Q}$ is orthogonal. By use of (11.16), show that (17.1) implies the requirement that

$$\mathrm{tr}[(\mathbf{Q} - \mathbf{1})\mathbf{T}(\mathbf{Q}^{T^*} - \mathbf{1})] > 0 \tag{17.2}$$

if $\mathbf{Q} \neq \mathbf{1}$. Prove that if (17.2) is to hold for all orthogonal $\mathbf{Q}$, it is necessary and sufficient that

$$t_1 > 0 \ , \ t_2 > 0 \ , \ t_3 > 0 \ , \tag{17.3}$$

while if (17.2) is to hold for all proper orthogonal $\mathbf{Q}$, or for all infinitesimal rotations, it is necessary and sufficient that

$$t_1 + t_2 > 0 \ , \quad t_2 + t_3 > 0 \ , \quad t_3 + t_1 > 0 \ . \tag{17.4}$$

In particular, (17.1) can never hold under pure rotations from a natural state.

The results of the foregoing exercise show that the inequality (17.1) is a frame-indifferent condition only in certain special states of stress, not even including an unstressed state. COLEMAN & NOLL proposed in a somewhat more special context to weaken the restriction by excluding rotations. That is, only deformation gradients $\mathbf{F}^*$ obtainable from $\mathbf{F}$ by a pure stretch are allowed: $\mathbf{F}^* = \mathbf{SF}$, where $\mathbf{S}$ is positive-definite and symmetric. The <u>generalized Coleman-Noll condition</u> (GCN) is the following <u>a priori</u> inequality for the response function $\mathfrak{h}$:

$$\mathrm{tr}\{[\mathfrak{h}(\mathbf{SF}) - \mathfrak{h}(\mathbf{F})][(\mathbf{SF})^T - \mathbf{F}^T]\} > 0 \tag{17.5}$$

if $\mathbf{S}$ is any positive-definite symmetric tensor other than $\mathbf{1}$.

We now consider the consequences that would follow, were this inequality laid down.

<u>Relation to Previous Results</u>.

First we consider an infinitesimal deformation from the reference configuration. Then $\mathbf{F} = \mathbf{1}$, $\mathbf{T}_o = \mathfrak{h}(\mathbf{1})$, and $\mathbf{S} = \mathbf{1} + \tilde{\mathbf{E}}$, where $\tilde{\mathbf{E}}$ is the infinitesimal strain tensor, and $\tilde{\mathbf{E}} \neq \mathbf{0}$. The condition (17.5) then becomes

$$\operatorname{tr}\{(\mathbf{T}_{\kappa} - \mathbf{T}_{o})\tilde{\mathbf{E}}\} = \operatorname{tr}\{\mathbf{A}_{o}[\tilde{\mathbf{E}}]\tilde{\mathbf{E}}\} > 0 \quad \text{if} \quad \tilde{\mathbf{E}} \neq \mathbf{0} , \tag{17.6}$$

where $\mathbf{A}_o$ is the elasticity $(11.20)_2$, evaluated at $\mathbf{F} = \mathbf{1}$. Thus the GCN condition implies that the specific work done in any non-vanishing infinitesimal pure strain is positive. Nothing is required of the stress work in non-pure strains.

<u>Exercise 17.2</u>. Using CAUCHY's law (14.12), show that in an infinitesimal pure strain

$$\mathbf{T}_{\kappa} = \mathbf{T}_o + \mathbf{T}_o \operatorname{tr} \tilde{\mathbf{E}} - \mathbf{T}_o \tilde{\mathbf{E}} + \mathbf{L}[\tilde{\mathbf{E}}] . \tag{17.7}$$

By (17.7) we may write (17.6) in the form

$$\operatorname{tr} \tilde{\mathbf{E}} \operatorname{tr}(\mathbf{T}_o \tilde{\mathbf{E}}) - \operatorname{tr}(\mathbf{T}_o \tilde{\mathbf{E}}^2) + \operatorname{tr}(\mathbf{L}[\tilde{\mathbf{E}}]\tilde{\mathbf{E}}) > 0$$

$$\text{if} \quad \tilde{\mathbf{E}} \neq 0 . \tag{17.8}$$

If the reference configuration is a natural state, $\mathbf{T}_o = \mathbf{0}$, and (17.8) reduces to the classical requirement (16.2). Thus for <u>infinitesimal deformation from a natural state, the GCN condition is equivalent to the classical requirement that the stress work in any non-rigid deformation be positive</u>. More generally, the elasticity $\mathbf{L}$ and the stress $\mathbf{T}_o$ must be such as to render the more complicated quadratic form on the left-hand side of (17.8) positive-definite. By (14.10) and the fact that $\mathbf{T}_o = \mathfrak{g}(\mathbf{1})$, this is a requirement upon the response function and its derivative at $\mathbf{F} = \mathbf{1}$.

Next, consider an arbitrary deformation of an <u>isotropic</u> elastic material from an undistorted state. Consider two deformation gradients $\mathbf{F}$ and $\mathbf{F}^*$ having the same rotation $\mathbf{R}$ and having also left stretch tensors $\mathbf{V}$ and $\mathbf{V}^*$ that commute:

$$\mathbf{F}^* = \mathbf{V}^*\mathbf{R} , \quad \mathbf{F} = \mathbf{V}\mathbf{R} , \quad \mathbf{V}\mathbf{V}^* = \mathbf{V}^*\mathbf{V} . \tag{17.9}$$

Then $\mathbf{F}^* = \mathbf{S}\mathbf{F}$, where $\mathbf{S} = \mathbf{V}^*\mathbf{V}^{-1}$. Consequently $\mathbf{F}^*$ and $\mathbf{F}$ furnish a particular pair of arguments to which (17.5) applies. In view of (11.40), namely,

$$\mathbf{T}_{\kappa} = (J\mathbf{T}\mathbf{V}^{-1})\mathbf{R} , \tag{11.40}$$

it follows that

$$(\underset{\sim}{T}^*_K - \underset{\sim}{T}_K)(\underset{\sim}{F}^{*T} - \underset{\sim}{F}^T) = (J^*\underset{\sim}{T}^*\underset{\sim}{V}^{*-1} - J\underset{\sim}{T}\underset{\sim}{V}^{-1})(\underset{\sim}{V}^* - \underset{\sim}{V}) . \tag{17.10}$$

Since $\underset{\sim}{V}$ and $\underset{\sim}{V}^*$ commute, they have a common orthonormal set of principal directions, with proper numbers v^*_i and v_i, and since the material is isotropic, these directions are principal directions also for $J^*\underset{\sim}{T}^*\underset{\sim}{V}^{*-1}$ and for $J\underset{\sim}{T}\underset{\sim}{V}^{-1}$, with corresponding proper numbers T^*_i and T_i. Hence

$$\mathrm{tr}[(\underset{\sim}{T}^*_K - \underset{\sim}{T}_K)(\underset{\sim}{F}^{*T} - \underset{\sim}{F}^T)] = \sum_{i=1}^{3} (T^*_i - T_i)(v^*_i - v_i) . \tag{17.11}$$

Comparison of (17.5) with (16.18) establishes the implication

$$\mathrm{GCN} \Rightarrow \mathrm{GCN}_o . \tag{17.12}$$

Thus the GCN condition includes and generalizes the GCN_o condition, the implications of which formed the subject to the last lecture.

Since a special pair of arguments was selected, we cannot expect to reverse the implication in (17.12).

Differential Form.

In the last lecture, the differential form defined by (16.21) was introduced, and among the results of Exercise 16.6 are two differential conditions: (1) For the GCN_o condition to hold, it is necessary that $\Sigma J_{ab}Z_aZ_b \geqq 0$ for all real non-vanishing Z_a, and, as a partial converse, (2) if $\Sigma J_{ab}Z_aZ_b > 0$ for all non-vanishing real Z_a, the GCN_o condition does hold. The stronger differential inequality was called the GCN^+_o condition. We now establish a corresponding pair of differential inequalities for the more general GCN condition.

Let $\underset{\sim}{D}$ be any non-zero symmetric tensor, and define $\bar{\underset{\sim}{F}}(\tau)$ by

$$\bar{\underset{\sim}{F}}(\tau) = (\underset{\sim}{1} + \tau\underset{\sim}{D})\underset{\sim}{F} , \tag{17.13}$$

where τ is a real parameter. Then $\bar{\underset{\sim}{F}}(0) = \underset{\sim}{F}$, and for small enough positive $|\tau|$ the tensor $\underset{\sim}{1} + \tau\underset{\sim}{D}$ is symmetric, positive-definite, and not $\underset{\sim}{1}$. Hence the pair $\underset{\sim}{F}$, $\bar{\underset{\sim}{F}}(\tau)$ qualify for the application of (17.1). Thus if we set

$$f(\tau) \equiv \mathrm{tr}\{(\underset{\sim}{D}\underset{\sim}{F})^T \underset{\sim}{\mathfrak{h}} [\bar{\underset{\sim}{F}}(\tau)]\} , \tag{17.14}$$

the GCN condition asserts that

$$\tau[f(\tau) - f(0)] > 0 . \tag{17.15}$$

Hence $f'(0) \geqq 0$. Now by $(11.20)_2$ and (17.13)

$$f'(\tau) = \mathrm{tr}\ \{(\underset{\sim}{D}\underset{\sim}{F})^T A[\overline{\underset{\sim}{F}}(\tau)][\underset{\sim}{D}\underset{\sim}{F}]\}\ , \tag{17.16}$$

so

$$J^{-1}f'(0) = J^{-1}\ \mathrm{tr}\{(\underset{\sim}{D}\underset{\sim}{F})^T A(\underset{\sim}{F})[\underset{\sim}{D}\underset{\sim}{F}]\},$$

$$= J^{-1}A_{k\ m}^{\ \alpha\ \beta}D^m_{\ q}F^q_{\ \beta}D^k_{\ r}F^r_{\ \alpha}\ . \tag{17.17}$$

Accordingly, if we set

$$B_{k\ m}^{\ r\ q} \equiv J^{-1}A_{k\ m}^{\ \alpha\ \beta}F^q_{\ \beta}F^r_{\ \alpha}\ , \tag{17.18}$$

the condition $f'(0) \geqq 0$ assumes the form

$$\beta[\underset{\sim}{D}, \underset{\sim}{D}] \equiv B^{kmpq}D_{km}D_{pq} \geqq 0 \tag{17.19}$$

for all symmetric tensors $\underset{\sim}{D}$. This is the required differential inequality implied by GCN.

The stronger inequality

$$\beta[\underset{\sim}{D}, \underset{\sim}{D}] \equiv B^{kmpq}D_{km}D_{pq} > 0 \tag{17.20}$$

if $\underset{\sim}{D}$ is a non-vanishing symmetric tensor is called the <u>GCN^+ condition</u>. It is natural to expect that

$$GCN^+ \Rightarrow GCN\ , \tag{17.21}$$

and this is true. To prove it, we assume given $\underset{\sim}{F}$ and a symmetric positive-definite tensor $\underset{\sim}{S}$, different from $\underset{\sim}{1}$. We put $\underset{\sim}{D} \equiv \underset{\sim}{S}-\underset{\sim}{1}$ and define $\underset{\sim}{F}(\tau)$ and $f(\tau)$ by (17.13) and (17.14).

<u>Exercise 17.3</u>. When $0 < \tau \leqq 1$, prove that $\underset{\sim}{1} + \tau\underset{\sim}{D}$ is invertible and $\underset{\sim}{D}(\underset{\sim}{1} + \tau\underset{\sim}{D})^{-1}$ is symmetric. Hence $\underset{\sim}{D}\underset{\sim}{F} = \underset{\sim}{D}(\underset{\sim}{1} + \tau\underset{\sim}{D})^{-1}\overline{\underset{\sim}{F}}$ and

$$f'(\tau) = \mathrm{tr}\ \{[\underset{\sim}{D}(\underset{\sim}{1} + \tau\underset{\sim}{D})^{-1}\overline{\underset{\sim}{F}}(\tau)]\ A[\overline{\underset{\sim}{F}}(\tau)][\underset{\sim}{D}(\underset{\sim}{1} + \tau\underset{\sim}{D})^{-1}\overline{\underset{\sim}{F}}(\tau)]\}. \tag{17.22}$$

The GCN^+ condition (17.20) applied at the argument $\overline{F}(\tau)$ and with $\underset{\sim}{D}$ replaced by $\underset{\sim}{D}(\underset{\sim}{1} + \tau\underset{\sim}{D})^{-1}$ is equivalent to the assertion that $f'(\tau) > 0$ when $0 < \tau \leqq 1$. Integration yields $f(1) - f(0) > 0$. By (17.14), this result is the GCN condition, so the implication (17.21) is established.

The GCN^+ Condition for Isotropic Materials.

For isotropic materials the GCN^+ condition may be given a simple statical interpretation. If we regard symmetric tensors as vectors in a 6-dimensional space, (17.20) asserts that β is a positive-definite quadratic form in such vectors.

Exercise 17.4. Prove that

$$B^{km}{}_{p}{}^{q} = T^{km}\delta^{q}_{p} - T^{kq}\delta^{m}_{p} + 2F^{km}{}_{ps}B^{qs} , \tag{17.23}$$

where F is the elasticity (11.43). By use of (11.44) show that the 6 x 6 matrix of β in principal components is composed of four 3 x 3 blocks as follows:

$$\|\beta\| = \left\| \begin{array}{c|c} \dfrac{v_a v_b J_{ab}}{v_1 v_2 v_3} & 0 \\ \hline 0 & \mathrm{diag}(A_r) \end{array} \right\| , \tag{17.24}$$

where J_{ab} is defined by (16.21) and where

$$A_1 = \frac{(t_2 - t_3)(v_2^2 + v_3^2)}{2(v_2^2 - v_3^2)} - \frac{1}{4}(t_2 + t_3), \ldots, \tag{17.25}$$

Hence

$$GCN^+ \Leftrightarrow GCN_0^+ \ \& \ (A_r > 0) \tag{17.26}$$

Prove that

$$E \Rightarrow A_r > 0 , \tag{17.27}$$

where the E-inequalities are defined by (16.13).

The result (17.26) shows that the GCN^+ condition is equivalent to two simpler ones. The GCN_o^+ inequality was discussed at length in the previous lecture. The inequalities $A_r > 0$ are harder to interpret, but (17.27) gives a simple sufficient condition for them to hold. This same condition, by (16.14), is sufficient also for the B-E inequalities.

LECTURE 18: WAVE PROPAGATION.

Hadamard's Lemma.

Let S be a part of the boundary of a region, which we shall denote by R_+, and let $\underset{\sim}{x}$ be a point on S. The field $\Psi(\underset{\sim}{x})$ is said to be smooth in R_+ if it is continuously differentiable in R_+, if for every point $\underset{\sim}{x}$ on S the fields $\Psi(\underset{\sim}{y})$ and $\partial_{\underset{\sim}{y}}\Psi(\underset{\sim}{y})$ approach limits $\overset{+}{\Psi}(\underset{\sim}{x})$ and $\overset{+}{\partial}_{\underset{\sim}{x}}\Psi(\underset{\sim}{x})$ as $\underset{\sim}{y} \to \underset{\sim}{x}$, and if $\overset{+}{\Psi}(\underset{\sim}{x})$ is differentiable on any path p lying on S. HADAMARD's lemma asserts that for a smooth field $\Psi(\underset{\sim}{x})$, the theorem of the total differential holds for the limit functions $\overset{+}{\Psi}$ and $\overset{+}{\partial}_{\underset{\sim}{x}}\Psi$. That is, if the path p is described by the parametric equation $\underset{\sim}{x} = \underset{\sim}{x}(\ell)$, then

$$\overset{+}{\Psi}{}'(\ell) = (\overset{+}{\partial}_{\underset{\sim}{x}}\Psi)\, \underset{\sim}{x}'(\ell) . \tag{18.1}$$

The field Ψ may be a scalar, vector, or tensor. The result (18.1) is written in a form appropriate for a scalar Ψ ; the notation needs to be modified in other cases.

Exercise 18.1. For the case when Ψ is a scalar, prove Hadamard's lemma.

Singular Surfaces.

Let the orientable surface S be a part of the common boundary separating two regions R_+ and R_-, in each of which Ψ is smooth. In this case, at a point $\underset{\sim}{x}$ on S the limits $\overset{+}{\Psi}$ and $\overset{-}{\Psi}$, and likewise $\overset{+}{\partial}_{\underset{\sim}{x}}\Psi$ and $\overset{-}{\partial}_{\underset{\sim}{x}}\Psi$, exist but need not be equal. The jumps of Ψ and $\partial_{\underset{\sim}{x}}\Psi$ are defined as the differences in these values:

$$[\Psi](\underset{\sim}{x}) \equiv [\Psi] \equiv \overset{+}{\Psi} - \overset{-}{\Psi} , \quad [\partial_{\underset{\sim}{x}}\Psi] \equiv \overset{+}{\partial}_{\underset{\sim}{x}}\Psi - \overset{-}{\partial}_{\underset{\sim}{x}}\Psi . \tag{18.2}$$

If one or both of these jumps is not zero, S is said to be singular with respect to Ψ.

A singular surface, then, is not merely one where some property of continuity or differentiability fails, since Ψ is required to be smooth on each side. The jumps possible across a singular surface are strictly limited in kind. Since $[\Psi](\underset{\sim}{x})$ a differentiable function of $\underset{\sim}{x}$ on S, applying Hadamard's lemma to Ψ^+ and Ψ^- and subtracting yields

$$[\Psi]'(x) = [\partial_{\underset{\sim}{x}}\Psi] \cdot \underset{\sim}{x}'(\ell) \ . \tag{18.3}$$

This is HADAMARD's <u>fundamental condition of compatibility</u>, which relates the jumps possible in Ψ and $\partial_{\underset{\sim}{x}} \Psi$.

An important corollary follows when Ψ is continuous: $[\Psi] = 0$. Then (18.3) yields

$$[\partial_{x}\Psi] \cdot \underset{\sim}{x}'(\ell) = 0 \tag{18.4}$$

for all paths on S. Since $\underset{\sim}{x}'(\ell)$ may be any vector tangent to S, (18.4) requires that there exist a quantity $a(\underset{\sim}{x})$, defined if $\underset{\sim}{x} \in S$, such that

$$[\partial_{\underset{\sim}{x}}\Psi] = a\underset{\sim}{n} \quad , \tag{18.5}$$

where $\underset{\sim}{n}$ is a vector normal to S. This result expresses <u>MAXWELL's theorem</u>: The jump of the gradient of a continuous field is normal to the singular surface. It is convenient to adopt a convention of sign; <u>e.g.</u>, let $\underset{\sim}{n}$ be the unit normal pointing from R_- into R_+. Then a, which is called the <u>amplitude</u> of the discontinuity, is uniquely determined. The form (18.5) is appropriate to a scalar field Ψ. If Ψ is a vector field, the amplitude a becomes a vector, $\underset{\sim}{a}$, and the right-hand side of (18.5) should be written as $\underset{\sim}{a} \otimes \underset{\sim}{n}$.

When $\underset{\sim}{a}$ is a vector, the singularity is called <u>longitudinal</u> if $\underset{\sim}{a} \parallel \underset{\sim}{n}$, <u>transverse</u> if $\underset{\sim}{a} \perp \underset{\sim}{n}$. In general, a singularity is neither.

<u>Singular Surfaces for a Motion</u>.

We consider the motion (1.11) with respect to a reference configuration $\underset{\sim}{\kappa}$:

$$\underset{\sim}{x} = \underset{\sim}{\chi}_{\underset{\sim}{\kappa}}(\underset{\sim}{X},t) \ , \tag{1.11}$$

and its derivatives $\dot{\underset{\sim}{x}}$, $\ddot{\underset{\sim}{x}}$, $\underset{\sim}{F}$, $\nabla\underset{\sim}{F}$, etc. The surface S in the reference configuration $\underset{\sim}{\kappa}$ is said to be a singular surface of nth order if it is singular with respect to some n^{th} derivative of $\underset{\sim}{\chi}$, but all derivatives of lower order exist and are continuous in a region containing S in its interior. The surface S is allowed to move in the reference configuration. Moreover, we consider only singular surfaces that persist throughout an interval of time. Thus they may be regarded as surfaces in a 4-dimensional space whose points are pairs $(\underset{\sim}{x},t)$, and Hadamard's condition (18.3) may be applied in that space.

Singular surfaces of orders 0 and 1 are called strong; shock waves, vortex sheets, tears, and welds are included. Singular surfaces of order 2 or greater are called weak. The mathematical theory of strong singular surfaces is rather intricate and will not be considered in these lectures.

Compatibility Conditions for Second-Order Singular Surfaces

Application of Maxwell's theorem (18.5) to second-order singular surfaces in $\underset{\sim}{\kappa}$ yields a formally simple relation in the reference configuration, with $\underset{\sim}{n}$ being the 4-dimensional unit normal. Separation into spatial and temporal parts, followed by transformation to the configuration $\underset{\sim}{\chi}$, yields

$$
\begin{aligned}
&[\nabla\underset{\sim}{F}] = \underset{\sim}{a} \otimes (\underset{\sim}{F}^T\underset{\sim}{n}) \otimes (\underset{\sim}{F}^T\underset{\sim}{n}), && [\dot{x}^k{}_{;\alpha\beta}] = a^k x^p{}_{,\alpha} x^q{}_{,\beta} n_p n_q ,\\
&[\dot{\underset{\sim}{F}}] = - U\underset{\sim}{a} \otimes (\underset{\sim}{F}^T\underset{\sim}{n}) , && [\dot{x}^k{}_{,\alpha}] = - U a^k x^p{}_{,\alpha} n_p , \qquad (18.6)\\
&[\ddot{\underset{\sim}{x}}] = U^2 \underset{\sim}{a} , && [\ddot{x}^k] = U^2 a^k ,
\end{aligned}
$$

where now $\underset{\sim}{n}$ is the unit normal to the present configuration of S, $\underset{\sim}{a}$ is a vector called the amplitude, and U is a scalar called the local speed of propagation.

The first of these equations, called HADAMARD's geometrical condition of compatibility, reflect the assumption that the discontinuity is spread out over a surface at the instant in question. The second two, called HADAMARD's kinematical conditions of compatibility, reflect the assumption that the singular surface instantaneously persists. If a point $\underset{\sim}{x}$ on the surface is moving with velocity $\underset{\sim}{v}$, and if the velocity of the particle instantaneously at $\underset{\sim}{x}$ is $\dot{\underset{\sim}{x}}$, then

$$U = (\underset{\sim}{v} - \dot{\underset{\sim}{x}}) \cdot \underset{\sim}{n} \; . \tag{18.7}$$

That is, U at a place and time is the normal speed of advance of the singular surface relative to the particle instantaneously situated at $\underset{\sim}{x}$. If $U \neq 0$, the singular surface <u>propagates</u> through the material and hence is called a <u>wave</u>. If $U = 0$ over an interval of time, the singular surface divides two portions of material. From $(18.6)_3$ it is clear that every second-order wave carries a non-zero jump of the acceleration. For this reason, such waves are called <u>acceleration waves</u>.

<u>Exercise 18.2</u>. Filling in the argument outlined above, prove (18.6). Prove also that the jump in the velocity gradient $\underset{\sim}{G}$ satisfies

$$[\underset{\sim}{G}] = - U \underset{\sim}{a} \otimes \underset{\sim}{n} \; , \tag{18.8}$$

and hence derive <u>WEINGARTEN's theorem</u>:

$$[\underset{\sim}{W}] = -U \underset{\sim}{a} \wedge \underset{\sim}{n} \; , \qquad [\operatorname{div} \dot{\underset{\sim}{x}}] = - U \underset{\sim}{a} \cdot \underset{\sim}{n} \; . \tag{18.9}$$

By use of WEINGARTEN's theorem, interpret the normal and tangential components of the amplitude $\underset{\sim}{a}$.

<u>Equation of Balance at a Singular Surface</u>.

In Lecture 3 we considered a general equation of balance:

$$\Big(\int_{p_{\underset{\sim}{\chi}}} \Psi \, dm\Big)^{\cdot} = \int_{\partial p_{\underset{\sim}{\chi}}} \underset{\sim}{E} \, \underset{\sim}{n} \, ds + \int_{p_{\underset{\sim}{\chi}}} s \, dm \; . \tag{3.13}$$

This equation may be applied also in the case when p contains or is divided by a weak singular surface S. If S is a singular surface also with respect to Ψ, as defined above, then the jump of Ψ is subjected to the following requirement, called <u>KOTCHINE's theorem</u>:

$$U[\rho\Psi] + [\underset{\sim}{E}] \cdot \underset{\sim}{n} = 0 \; . \tag{18.10}$$

and conversely, if (18.10) holds at each point of S and (3.15) holds at interior points of a region containing S in its interior, then (3.13) holds in that region.

Exercise 18.3. Under the assumption that s is bounded in a region containing S, and bearing in mind the restrictions on Ψ already stated in the definition of "singular surface", prove KOTCHINE's theorem.

Applying KOTCHINE's theorem to the equation of balance of momentum, namely, (3.12), we put $\Psi = \dot{\underset{\sim}{x}}$ and $\underset{\sim}{E} = \underset{\sim}{T}$ and obtain POISSON's condition:

$$[\underset{\sim}{T}]\, \underset{\sim}{n} = \underset{\sim}{0} \;. \tag{18.11}$$

According to POISSON's condition, at a weak singular surface the balance of momentum requires only that the stress vector be continuous.

Acceleration Waves in Elasticity.

To consider acceleration waves in elasticity, we write the constitutive equation in the form

$$\underset{\sim}{T}_{\underset{\sim}{\kappa}} = \underset{\sim}{\mathfrak{h}}\,(\underset{\sim}{F},\underset{\sim}{X}) \tag{11.15}$$

and assume that $\underset{\sim}{\mathfrak{h}}$ is continuously differentiable with respect to each of its arguments. At a weak singular surface POISSON's condition (18.11) is then satisfied. On each side of the singular surface, the deformation satisfies the differential equation (11.21), viz

$$A_{p\ m}^{\ \alpha\ \beta} x^m{}_{,\alpha;\beta} + q_p + \rho_{\underset{\sim}{\kappa}} b_p = \rho_{\underset{\sim}{\kappa}} \ddot{x}_p \;, \tag{11.21}$$

where

$$A_{p\ m}^{\ \alpha\ \beta} \equiv \partial_{x^m{}_{,\beta}}\, \mathfrak{h}_p^{\ \alpha} \;, \quad q_p \equiv \partial_{X^\alpha}\, \mathfrak{h}_p^{\ \alpha} \;. \tag{11.20$_{1,3}$}$$

From our hypothesis regarding $\underset{\sim}{\mathfrak{h}}$ it follows that A and q as just defined are continuous across S. We shall assume that $\underset{\sim}{b}$ also is continuous. At a point on S, we may approach from the + side and obtain then in the limit

$$A_p{}^{\alpha}{}_m{}^{\beta}(x^m{}_{,\alpha;\beta})^+ + q_p + \rho_{\underset{\sim}{\kappa}} b_p = \rho_{\underset{\sim}{\kappa}} \ddot{x}_p^+ , \tag{18.12}$$

while approach from the - side yields the limit relation

$$A_p{}^{\alpha}{}_m{}^{\beta}(x^m{}_{,\alpha;\beta})^- + q_p + \rho_{\underset{\sim}{\kappa}} b_p = \rho_{\underset{\sim}{\kappa}} \ddot{x}_p^- . \tag{18.13}$$

Taking the difference yields

$$A_p{}^{\alpha}{}_m{}^{\beta}[x^m{}_{,\alpha;\beta}] = \rho_{\underset{\sim}{\kappa}}[\ddot{x}_p] . \tag{18.14}$$

In this result we substitute Hadamard's conditions $(18.6)_{1,3}$ in their component forms. If we set

$$Q_{pm}(\underset{\sim}{n}) \equiv \frac{\rho}{\rho_{\underset{\sim}{\kappa}}} A_p{}^{\alpha}{}_m{}^{\beta} x^r{}_{,\alpha} x^s{}_{,\beta} n_r n_s , \tag{18.15}$$

then the result of the substitution is

$$(\underset{\sim}{Q}(\underset{\sim}{n}) - \rho U^2 \underset{\sim}{1})\underset{\sim}{a} = \underset{\sim}{0} . \tag{18.16}$$

$\underset{\sim}{Q}(\underset{\sim}{n})$ is called the <u>acoustic tensor</u> in the direction $\underset{\sim}{n}$ for the elastic material at $\underset{\sim}{X}$ when subjected to the deformation $\underset{\sim}{F}$. According to (18.16), which is called the <u>propagation condition</u>, any amplitude $\underset{\sim}{a}$ of a second-order singular surface with normal $\underset{\sim}{n}$ must be a right proper vector of $\underset{\sim}{Q}(\underset{\sim}{n})$, and its speed of propagation U is such that ρU^2 is the corresponding proper number. The directions corresponding to the proper numbers $\underset{\sim}{a}$ are called the <u>acoustic axes</u> for waves traveling in the direction $\underset{\sim}{n}$ at $\underset{\sim}{x}$, t. In the generality here maintained, little can be said about the number and nature of the acoustic axes and the corresponding speeds of propagation. In the special case when $\underset{\sim}{Q}(\underset{\sim}{n})$ is symmetric,

$$\underset{\sim}{Q}(\underset{\sim}{n}) = \underset{\sim}{Q}(\underset{\sim}{n})^T , \tag{18.17}$$

application of a standard theorem tells us that at least one orthogonal triple of real acoustic axes for the direction $\underset{\sim}{n}$ exists, and that the corresponding squared speeds U^2 are real. It remains possible, however, that the speeds U may be purely imaginary, so that the corresponding singular surfaces do not exist. The foregoing remarks, summarizing the implications of the propagation condition (18.16), constitute the <u>FRESNEL-HADAMARD theorem.</u>

<u>Exercise 18.4</u>. Using the general theory presented in Lecture 14, prove that the speeds and amplitudes of free plane sinusoidal oscillations about a homogeneous configuration satisfy the propagation condition (18.6).

<u>Exercise 18.5</u>. Prove that the acoustic tensor $\underset{\sim}{Q}(\underset{\sim}{n})$ for the deformation gradient $\underset{\sim}{F}$ is symmetric for <u>every</u> $\underset{\sim}{n}$ if and only if the operator A is self-adjoint at the argument $\underset{\sim}{F}$:

$$A_{p}{}^{\alpha}{}_{m}{}^{\beta} = A_{m}{}^{\beta}{}_{p}{}^{\alpha} . \tag{18.18}$$

Prove that if (18.18) holds for all $\underset{\sim}{F}$, there exists a potential function $\sigma(\underset{\sim}{F})$ for the Piola-Kirchhoff stress:

$$\underset{\sim}{T}_{\underset{\sim}{\kappa}} = \rho_{\underset{\sim}{\kappa}} \partial_{\underset{\sim}{F}} \sigma(\underset{\sim}{F}) , \tag{18.19}$$

and conversely.

An elastic material whose stress relation (11.15) has the special form (18.19) is called <u>hyperelastic</u>, and $\sigma(\underset{\sim}{F})$ is called the <u>stored-energy function</u>. In some of the following lectures I shall present other properties of hyperelastic materials, but for the time being there is no need to limit attention to them.

<u>Weak Singular Surfaces in General</u>.

The Fresnel-Hadamard theorem applies not only to acceleration waves but to all weak surfaces of any order.

Exercise 18.6. Prove that the geometrical and kinematical conditions of compatibility for a third-order singular surface are

$$[x^m{}_{,\alpha;\beta\gamma}] = a^m x^r{}_{,\alpha} n_r x^s{}_{,\beta} n_s x^u{}_{,\gamma} n_u ,$$

$$[\dddot{x}_p] = -U^3 a_p . \qquad (18.20)$$

Hence show that the amplitudes and speeds of third-order waves satisfy the propagation condition (18.16).

Since the propagation condition (18.16) expresses properties common to so many kinds of disturbances, it is customary to call the theory based upon it acoustics and to refer, loosely, to any wave described by it as a sound wave. The speeds U are often called speeds of sound.

The S-E and H Conditions.

If it is known that $\underset{\sim}{a}$ is a possible amplitude, the corresponding speeds may be calculated at once from (18.16):

$$\rho U^2 = \frac{\underset{\sim}{a} \cdot \underset{\sim}{Q}(\underset{\sim}{n})\underset{\sim}{a}}{a^2} . \qquad (18.21)$$

Accordingly, the condition

$$\underset{\sim}{a} \cdot \underset{\sim}{Q}(\underset{\sim}{n})\underset{\sim}{a} > 0 \qquad (18.22)$$

for all real vectors $\underset{\sim}{n}$ and $\underset{\sim}{a}$ is sufficient that the two speeds U corresponding to any real amplitude be real and non-vanishing. This condition, asserting that the quadratic form based on the symmetric part of $\underset{\sim}{Q}(\underset{\sim}{n})$ is positive definite, is called the S-E condition. The weaker condition

$$\underset{\sim}{m} \cdot \underset{\sim}{Q}(\underset{\sim}{n})\underset{\sim}{m} \geqq 0 \qquad (18.23)$$

is called HADAMARD's condition (H). When $\underset{\sim}{Q}(\underset{\sim}{n})$ is symmetric all its proper vectors are real (or trivially, purely imaginary), and the S-E condition asserts that for each direction of propagation there exists at least one orthogonal set of acoustic axes, and a singular surface corresponding to any acoustic axis must propagate. Roughly speaking, the S-E condition insures the existence of at least three possible independent amplitudes for sound waves. The H-condition is weaker in that $U = 0$ is not excluded, so non-propagating singular surfaces are allowed.

If we return to the definition (18.1), we can write the S-E condition in the form

$$A_{k\ m}^{\alpha\ \beta} a^k a^m b_\alpha b_\beta > 0 \tag{18.24}$$

for arbitrary vectors non-vanishing $\underset{\sim}{a}$ and $\underset{\sim}{b}$, or, in direct notation.

$$\mathrm{tr}\{(A(\underset{\sim}{F})[\underset{\sim}{a} \otimes \underset{\sim}{b}])(\underset{\sim}{a} \otimes \underset{\sim}{b})^T\} > 0 \ . \tag{18.25}$$

This condition is familiar in analysis. When it is satisfied, the operator $A(\underset{\sim}{F})$ is said to be <u>strongly elliptic</u>. The H-condition is the slightly weaker inequality resulting when ">" is replaced by "$\geqq$" in (18.24) and (18.25).

<u>Exercise 18.7</u>. Using (17.18), prove that the S-E and H conditions may be put into the forms

$$B^{kmpq} a_k b_m a_p a_q > 0 \quad \text{and} \quad \geqq 0 \ , \tag{18.26}$$

respectively.

<u>Comparison of the S-E and GCN conditions</u>.

If we compare the S-E condition with the GCN^+ condition (17.20), we see that both may be expressed in terms of a common quadratic form:

$$\mathrm{tr}\{A[\underset{\sim}{G}]\underset{\sim}{G}^T\} > 0 \tag{18.27}$$

where GCN^+ asserts the inequality for all <u>symmetric</u> $\underset{\sim}{G}$, while S-E asserts it for all $\underset{\sim}{G}$ of <u>rank 1</u>, <u>viz</u>, of the form $\underset{\sim}{G} = \underset{\sim}{a} \otimes \underset{\sim}{b}$. Alternatively, these two conditions may be written in the form

$$B[\underset{\sim}{G}, \underset{\sim}{G}] > 0 \ , \tag{18.28}$$

with the same difference of tensors $\underset{\sim}{G}$ for which inequality is asserted.

If ">" is replaced by "$\geqq$", the H-condition results if $\underset{\sim}{G} = \underset{\sim}{a} \otimes \underset{\sim}{b}$, while restricting $\underset{\sim}{G}$ to be symmetric yields the inequality (17.19), shown in the last lecture to be necessary but not sufficient for the GCN condition to hold.

These results suggest, and it is possible to prove by examples, that S-E and GCN are distinct conditions. In general, neither implies the other.

To lay down (18.28) or (18.27) for general $\underset{\sim}{G}$ would of course imply that both GCN^+ and S-E hold. As shown in somewhat different terms in the preceding lecture, such a requirement is forbidden because it is not frame-indifferent except in a particular class of stress fields.

Exercise 18.8. Prove that a necessary and sufficient condition that $\underset{\sim}{T} = \underset{\sim}{0}$ is

$$B^{kmpq} = B^{mkpq} = B^{kmqp} . \tag{18.29}$$

Hence prove that in a natural state

$$GCN^+ \Rightarrow \text{S-E} , \tag{18.30}$$

$$GCN \Rightarrow H .$$

In the last lecture the GCN condition was shown to reduce for infinitesimal deformations from a natural state to the standard a priori inequality of classical elasticity theory. The result (18.30) expresses a standard theorem of that theory: For any wave normal, there exists at least one set of real orthogonal acoustic axes, and the corresponding squared speeds of sound are positive.

Acceleration Waves in Isotropic Materials

We have derived already a special form of Cauchy's first law of motion for homogeneous isotropic materials:

$$F^{km}{}_{pq} B^{pq}{}_{,m} + \rho b^k = \rho \ddot{x}^k , \tag{11.42}$$

where F is defined by (11.43). Either directly from this equation, or by substitution in (18.16), it is easy to show that the acoustic tensor $\underset{\sim}{Q}(\underset{\sim}{n})$ for an isotropic material is given by

$$Q^k{}_m(\underset{\sim}{n}) = 2F^{kp}{}_{mq} B^{qr} n_r n_p . \tag{18.31}$$

In a principal frame, of course

$$[\underset{\sim}{B}] = \operatorname{diag}\,(v_1^2,\ v_2^2,\ v_3^2)\ . \tag{18.32}$$

Let the unit vector $\underset{\sim}{n}$ have the principal components $\cos\theta_1$, $\cos\theta_2$, $\cos\theta_3$. Using (18.31), (18.32), and the formulae (11.44) for the principal components of F, it is easy to calculate the principal components of $\underset{\sim}{Q}(\underset{\sim}{n})$:

$$Q\langle 11\rangle(\underset{\sim}{n}) = (\partial_{\log v_1} t_1)\cos^2\theta_1 + \frac{t_1-t_2}{v_1^2-v_2^2}\, v_2^2 \cos^2\theta_2 + \frac{t_1-t_3}{v_1^2-v_3^2}\, v_3^2 \cos^2\theta_3,$$

$$Q\langle 12\rangle(\underset{\sim}{n}) = \left[\partial_{\log v_2} t_1 + \frac{t_1-t_2}{v_1^2-v_2^2}\, v_1^2\right] \cos\theta_1 \cos\theta_2, \tag{18.33}$$

A <u>principal wave</u> is a wave normal to a principal axis of stretch (and hence also, in an isotropic material, to a principal axis of stress). For example, we may take $\underset{\sim}{n}$ as the unit vector $\underset{\sim}{n}_1$ along the axis numbered 1, so it has components (1, 0, 0). Then (18.33) yields

$$\begin{aligned} Q\langle 11\rangle(\underset{\sim}{n}_1) &= \partial_{\log v_1} t_1\ , \\ Q\langle 22\rangle(\underset{\sim}{n}_1) &= v_1^2\, \frac{t_1-t_2}{v_1^2-v_2^2}\ , \\ Q\langle 12\rangle(\underset{\sim}{n}_1) &= 0\ ,\ \ldots \end{aligned} \tag{18.34}$$

Since $[\underset{\sim}{Q}(\underset{\sim}{n}_1)]$ is diagonal in this frame, we have shown that <u>the principal axes of stretch and stress are also the acoustic axes for principal waves.</u> Since the principal axes of stretch are orthogonal, it follows that <u>every principal wave in an isotropic material is either longitudinal or transverse</u>. The speeds of these waves may be calculated at once from (18.27) and (18.34).

First, take $\underset{\sim}{n}_1$ itself as amplitude, and write U_{11} for the speed of a corresponding longitudinal wave. Then

$$\begin{aligned} \rho U_{11}^2 &= \underset{\sim}{n}_1 \cdot \underset{\sim}{Q}(\underset{\sim}{n}_1)\underset{\sim}{n}_1\ , \\ &= Q\langle 11\rangle(\underset{\sim}{n}_1)\ , \\ &= \partial_{\log v_1} t_1\ . \end{aligned} \tag{18.35}$$

Second, let the amplitude be a transverse unit vector $\mathbf{n}_2$, say with components (0, 1, 0), and write U_{12} for the speed of a corresponding transverse wave. Then

$$
\begin{aligned}
\rho U_{12}^2 &= \mathbf{n}_2 \cdot \mathbf{Q}(\mathbf{n}_1)\mathbf{n}_2 , \\
&= Q^{\langle 12\rangle}(\mathbf{n}_1) , \qquad (18.36) \\
&= v_1^2 \frac{t_1 - t_2}{v_1^2 - v_2^2} .
\end{aligned}
$$

These formulae express the principal wave speeds directly in terms of "tangent moduli" of extension along the three principal axes and "secant moduli" of shear in the three planes perpendicular to them. In general, there are 9 distinct squared wave speeds U_{ab}^2.

Exercise 18.9. In a state of dilatation with stretch v from a natural state, subject to a pressure $p = p(\rho)$, prove that the 9 wave speeds reduce to 2. Writing $U_\perp$ for the speed of a transverse wave, $U_{||}$ for the speed of a longitudinal wave, prove that

$$
\begin{aligned}
U_\perp^2 &= \frac{v^5}{\rho_{\kappa}} (\aleph_1 + 2v^2 \aleph_2) , \\
& \qquad\qquad\qquad\qquad\qquad\qquad (18.37) \\
U_{||}^2 &= \frac{4}{3} \frac{v}{\rho_{\kappa}} (\aleph_1 + 2v^2 \aleph_2) + p'(\rho) ,
\end{aligned}
$$

where the $\aleph_r$ are the response coefficients in the stress relation (11.36). By specializing these results to the case of infinitesimal deformation from a natural state, derive the POISSON - CHRISTOFFEL theorem:

$$
U_\perp^2 = \frac{\mu}{\rho} , \quad U_{||}^2 = \frac{\lambda + 2\mu}{\rho} . \qquad (18.38)
$$

Show also that for elastic fluids

$$
U_{||}^2 = p'(\rho) \qquad (18.39)
$$

and that more generally

$$
U_{||}^2 = \frac{4}{3} U_\perp^2 + p'(\rho) . \qquad (18.40)
$$

Hence show that in an elastic material in which transverse waves may propagate, the speed of longitudinal waves in a body subject to hydrostatic pressure is always greater than would follow from the hydrodynamics of a fluid with the same pressure function.

No simple static interpretation for the S-E condition is known, but some static consequences of it are easy to obtain.

Exercise 18.10. Prove that for isotropic materials

$$\text{S-E} \Rightarrow \text{T-E}^{+} \;\;\&\;\; \text{B-E} . \qquad (18.41).$$

LECTURE 19: INFINITESIMAL STABILITY AND UNIQUENESS

Kirchhoff's Identity.

Uniqueness is proved in the classical infinitesimal theory by means of an identity derived by Kirchhoff. The identity itself is easily generalized to the finite theory. We begin with the differential equations of equilibrium which follow from $(11.14)_1$, *viz*,

$$\operatorname{Div} \underset{\sim}{T}_{\kappa} + \rho_{\kappa} \underset{\sim}{b} = \underset{\sim}{0} , \tag{19.1}$$

where

$$\underset{\sim}{T}_{\kappa} = \underset{\sim}{\mathfrak{h}}(\underset{\sim}{F}) . \tag{11.15}$$

On the boundary $\partial \mathcal{B}_{\kappa}$ of the reference configuration $\mathcal{B}_{\kappa}$ of a body $\mathcal{B}$, let the corresponding reference tractions and positions after deformation be denoted by $\underset{\sim}{t}_{\kappa}$ and $\underset{\sim}{x}$, as usual. Now let a second configuration of the body $\mathcal{B}$, referred to $\mathcal{B}_{\kappa}$, be denoted by $\underset{\sim}{x}^* = \underset{\sim}{\chi}^*(\underset{\sim}{X})$, subject to body force $\underset{\sim}{b}^*$, with boundary values $\underset{\sim}{t}^*_{\kappa}$ and $\underset{\sim}{x}^*$, and with Piola stress $\underset{\sim}{T}^*_{\kappa} = \underset{\sim}{\mathfrak{h}}(\underset{\sim}{F}^*)$. Then of course

$$\operatorname{Div} \underset{\sim}{T}^*_{\kappa} + \rho_{\kappa} \underset{\sim}{b}^* = \underset{\sim}{0} . \tag{19.2}$$

Exercise 19.1. Derive Kirchhoff's identity:

$$\int_{\mathcal{B}_{\kappa}} \rho_{\kappa} (\underset{\sim}{b}^* - \underset{\sim}{b})\cdot(\underset{\sim}{x}^* - \underset{\sim}{x})dv + \int_{\partial \mathcal{B}_{\kappa}} (\underset{\sim}{x}^* - \underset{\sim}{x})\cdot(\underset{\sim}{t}^*_{\kappa} - \underset{\sim}{t}_{\kappa})ds \tag{19.3}$$

$$= \int_{\mathcal{B}_{\kappa}} \operatorname{tr}\{(\underset{\sim}{T}^*_{\kappa} - \underset{\sim}{T}_{\kappa})(\underset{\sim}{F}^{*T} - \underset{\sim}{F}^{T})\}dv .$$

If both of the configurations χ and χ^* correspond to the same body force, then $b^* = b$ in $\mathcal{B}_\kappa$, and if both correspond to the same mixed boundary conditions of place and traction, then on $\partial\mathcal{B}_\kappa$ either $x^* = x$ or $t^*_\kappa = t_\kappa$. Thus the left-hand side of (19.3) vanishes; hence

$$0 = \int_{\mathcal{B}_\kappa} \operatorname{tr} \{(T^*_\kappa - T_\kappa)(F^{*T} - F^T)\} dv . \tag{19.4}$$

In the infinitesimal theory, conditions such as to make the integrand positive unless F^* and F differ by a rotation are laid down a priori. It follows that the two deformation fields differ by at most a rotation.

Exercise 19.2. Using the a priori inequality (16.2), verify in detail the assertions made in the last two sentences.

In Lecture 17 we have considered a universal requirement of sign for the integrand,

$$\operatorname{tr}\{(T^*_\kappa - T_\kappa)(F^{*T} - F^T)\} > 0 \tag{17.1}$$

if $F^* \neq F$, and have rejected it as not being frame-indifferent. The arguments given there in objection refer only to deformations differing by a pure rotation: $F^* = QF$, where Q is orthogonal. We can circumvent them by disallowing such pairs of deformations, and also by requiring inequality only in mean rather than at each point. We shall say that a configuration $\mathcal{B}_\chi$ of an elastic body $\mathcal{B}$ with stress relation (11.15) is stable with respect to κ if

$$\int_{\mathcal{B}_\kappa} \operatorname{tr}\{[\mathfrak{h}(F^*) - \mathfrak{h}(F)][F^{*T} - F^T]\} dv \geqq 0 \tag{19.5}$$

for all pairs F, F^*; superstable with respect to κ if

$$\int_{\mathcal{B}_\kappa} \operatorname{tr}\{[\mathfrak{h}(F^*) - \mathfrak{h}(F)][F^{*T} - F^T]\} dv > 0 \tag{19.6}$$

for all F^* which do not differ from F by a pure rotation.

Despite their formal similarity, stability and superstability are distinct concepts. Stability does not imply superstability, since "=" for at least some pairs of distinct deformations $\underset{\sim}{F}$ and $\underset{\sim}{F}^*$ is allowed in (19.4). Superstability does not imply stability, since (19.5) is an assertion about all pairs $\underset{\sim}{F}$, $\underset{\sim}{F}^*$, while (19.6) imposes no condition at all on pairs that differ by a pure rotation.

Little can be done with these concepts for general deformations.

Infinitesimal Stability.

When only infinitesimal deformations are allowed, the foregoing considerations become less restrictive and more useful. Let us take as reference the configuration in which the stability of $\mathcal{B}$ is to be considered, say $\underset{\sim}{\kappa}$. For infinitesimal deformations (Lecture 14),

$$\underset{\sim}{T}_{\underset{\sim}{\kappa}} = \underset{\sim}{T}_0 + A_0[\underset{\sim}{H}] , \tag{19.7}$$

where A_0 is the elasticity $(11.19)_2$ evaluated at $\underset{\sim}{F} = \underset{\sim}{1}$ and where $\underset{\sim}{H}$ is the gradient of the superimposed infinitesimal deformation, so that $\underset{\sim}{F} = \underset{\sim}{H} + \underset{\sim}{1}$ (cf. (14.7)). Hence

$$\begin{aligned} \operatorname{tr}\{(\underset{\sim}{T}_{\underset{\sim}{\kappa}} - \underset{\sim}{T}_0)(\underset{\sim}{F}^{*T} - \underset{\sim}{F}^T)\} &= \operatorname{tr}\{A_0[\underset{\sim}{H}]\underset{\sim}{H}^T\} \\ &= A_{0k\ m}^{\ \alpha\ \beta}\, u^k_{\ ,\alpha} u^m_{\ ,\beta} , \end{aligned} \tag{19.8}$$

where $\underset{\sim}{u}$ is the infinitesimal displacement vector.

If

$$\int_{\mathcal{B}_{\underset{\sim}{\kappa}}} \operatorname{tr}\{A_0[\underset{\sim}{H}]\underset{\sim}{H}^T\}dv \geqq 0 \tag{19.9}$$

for the gradient $\underset{\sim}{H}$ of <u>any</u> displacement field $\underset{\sim}{u}$ that is compatible with assigned mixed boundary conditions of place and traction, the configuration $\mathcal{B}_{\underset{\sim}{\kappa}}$ is said to be <u>infinitesimally stable</u> for the boundary conditions considered. In such a configuration, then, the total work of the tractions required to effect a deformation compatible with the boundary conditions is at least as great as that which would be required, could the same deformation be effected under dead loading.

Clearly

$$\text{stability} \Rightarrow \text{infinitesimal stability}, \tag{19.10}$$

but in general the converse is false.

Similarly, if

$$\int_{\mathcal{B}_\kappa} \operatorname{tr}\{A_0[\tilde{H}]\tilde{H}^T\}dv > 0 \tag{19.11}$$

for all $\tilde{H}$ that are not pure rotations, i.e., for all $\tilde{H}$ that are not skew, the configuration is infinitesimally superstable.

In an infinitesimally superstable configuration, (19.11) holds. Consequently (19.4) is contradicted, unless, perhaps, $\tilde{F}^*$ and $\tilde{F}$ differ by a gradient which is also a skew tensor. As is shown in works in the infinitesimal theory, such a tensor is necessarily constant and hence represents a rigid infinitesimal rotation. Thus

$$\text{infinitesimal superstability} \Longrightarrow \text{static uniqueness to within infinitesimal rotation}, \tag{19.12}$$

where now "uniqueness" refers to the solution of the mixed boundary-value problem for infinitesimal displacements from the configuration from which superstability with respect to these same boundary values is asserted. This theorem is due in principle of ERICKSEN & TOUPIN.

Thus a configuration for which uniqueness of a static boundary-value problem fails to hold cannot be infinitesimally superstable. As we have said, however, it may be stable.

Exercise 19.3. Show that the a priori inequality (16.2) suffices that a natural state be both infinitesimally stable and infinitesimally superstable. Hence derive the classical uniqueness theorems of KIRCHHOFF and F. NEUMANN.

Exercise 19.4. Prove that if $\underset{\sim}{b} = \underset{\sim}{0}$, the differential equations of infinitesimal deformation are satisfied by sinusoidal oscillations: $\underset{\sim}{u}(\underset{\sim}{x}, t) = \underset{\sim}{U}(\underset{\sim}{x}) \sin\omega t$, where $\omega/2\pi$ is the frequency. Such oscillations are said to be free if

$$\int_{\partial\mathcal{B}_{\underset{\sim}{\kappa}}} \underset{\sim}{U}\cdot(\underset{\sim}{T}_{\underset{\sim}{\kappa}} - \underset{\sim}{T}_0)\underset{\sim}{n}_{\underset{\sim}{\kappa}}\, ds = 0 , \tag{19.13}$$

e.g., if in fact $\partial\mathcal{B}_{\underset{\sim}{\kappa}}$ is free of traction. Prove that if $\mathcal{B}_{\underset{\sim}{\kappa}}$ is infinitesimally stable, in free oscillation $\omega^2 \geqq 0$, and if $\mathcal{B}_{\underset{\sim}{\kappa}}$ is infinitesimally superstable, $\omega^2 > 0$.

Hadamard's Fundamental Theorem.

The conditions of infinitesimal stability and superstability are of integral form. It is natural to try to relate them to local inequalities.

In Lecture 17 we considered the GCN^+ condition (17.20). If we use subscript 0 to denote evaluation at $\underset{\sim}{F} = \underset{\sim}{1}$, by (17.18) we have $A_0 = B_0$. Thus the GCN^+ condition (17.20) may be written in the form

$$\mathrm{tr}\{A_0[\underset{\sim}{D}]\underset{\sim}{D}\} > 0 \tag{19.14}$$

for all symmetric and non-vanishing $\underset{\sim}{D}$. This condition cannot guarantee stability or superstability, since it gives no information at all regarding $\underset{\sim}{H}$ which are not symmetric. It does yield a restricted uniqueness theorem, viz,

$GCN^+ \Rightarrow$ no mixed static boundary-value problem can have two solutions differing by an infinitesimal pure strain. (19.15)

The proof is immediate from (19.3).

A major result of converse type is expressed by HADAMARD's fundamental theorem on stability:

$$\text{infinitesimal stability} \Rightarrow \text{H}. \tag{19.16}$$

Here H stands for HADAMARD's inequality (18.24), or

$$\mathrm{tr}\{A_0[\underset{\sim}{a} \otimes \underset{\sim}{b}](\underset{\sim}{a} \otimes \underset{\sim}{b})^T\} \geqq 0 \qquad (19.17)$$

for all vectors $\underset{\sim}{a}$ and $\underset{\sim}{b}$. This remarkable theorem asserts that in order for $\mathcal{B}_{\underset{\sim}{\kappa}}$ to be infinitesimally stable for any one mixed boundary-value problem, the local inequality (19.17) must hold at each point.

The proof of this theorem is too difficult to give here.

In view of the results on wave propagation given in Lecture 18, HADAMARD's theorem implies that in a configuration infinitesimally stable for any one mixed boundary-value problem, all squared wave-speeds are non-negative. When the acoustic tensor is symmetric, so that the acoustic axes are real and orthogonal, it follows that at every point of a stable configuration there exist for any given wave normal at least three mutually orthogonal amplitudes with real speeds of propagation. It is possible that one or more of those speeds may vanish.

Homogeneous Configurations

In general, A_o is a function of position in $\mathcal{B}_{\underset{\sim}{\kappa}}$. If $\underset{\sim}{\kappa}$ is a homogeneous configuration, of course A_o is a constant, and HADAMARD's inequality (19.17) becomes a single requirement rather than an infinity of them. In this case, a converse theorem has been proved by VAN HOVE, namely,

$$\text{H} \Longrightarrow \begin{array}{l}\text{infinitesimal stability}\\ \text{for the problem of place.}\end{array} \qquad (19.18)$$

Likewise

$$\text{S-E} \Longrightarrow \begin{array}{l}\text{infinitesimal superstability}\\ \text{for the problem of place.}\end{array} \qquad (19.19)$$

The proof is too difficult to include here.

For strongly elliptic systems with constant coefficients, BROWDER has proved that

$$\text{uniqueness} \Rightarrow \text{existence}, \qquad (19.20)$$

again for the problem of place. This result applies, in particular, to infinitesimal deformations from a homogeneous configuration.

Thus we have the following chain of theorems for the boundary-value problem of place for infinitesimal deformation from a homogeneous configuration:

$$\text{S-E} \Rightarrow \text{superstability} \Rightarrow \text{uniqueness} \Rightarrow \text{existence.} \qquad (19.21)$$

(Van Hove) (Ericksen & Toupin) (Browder)

At each step, of course, requirements of smoothness are laid down. These are not detailed here, since we give no more than a schema of the results. The existence and uniqueness theorem for the boundary-value problem of place in the classical infinitesimal theory is included in (19.21), since the classical a priori inequality implies S-E, as we saw from a more general standpoint in Lecture 18.

The Effect of Infinitesimal Rotations.

Exercise 19.5. Prove that the condition of infinitesimal stability (19.10) may be put in the following alternative form:

$$\int_{\mathcal{B}_{\underset{\sim}{K}}} \{\mathrm{tr}[\underset{\sim}{T}_0(\underset{\sim}{\tilde{E}}\underset{\sim}{\tilde{R}} + \underset{\sim}{\tilde{R}}\underset{\sim}{\tilde{E}} + \underset{\sim}{\tilde{R}}^T\underset{\sim}{\tilde{R}})] + \beta(\underset{\sim}{\tilde{E}},\underset{\sim}{\tilde{E}}]\}dv \geqq 0 \ , \qquad (19.22)$$

where $\underset{\sim}{\tilde{E}}$ and $\underset{\sim}{\tilde{R}}$ are the tensors of infinitesimal strain and rotation, and where β is defined by (17.19).

In a natural state, $\underset{\sim}{T}_0 = \underset{\sim}{0}$, and by (17.19), the GCN condition suffices for infinitesimal stability. In stressed states, no such implication holds. Consider, for example, a pure rotation: For stability it is necessary that

$$\int_{\mathcal{B}_{\underset{\sim}{K}}} \mathrm{tr}(\underset{\sim}{\tilde{R}}\ \underset{\sim}{T}_0\ \underset{\sim}{\tilde{R}}^T)dv \geqq 0 \ . \qquad (19.23)$$

Thus if

$$\mathrm{tr}(\underset{\sim}{\tilde{R}}\ \underset{\sim}{T}_0\ \underset{\sim}{\tilde{R}}^T) \ < 0 \ , \qquad (19.24)$$

the configuration cannot be infinitesimally stable, for any material. This condition with the inequality reversed has already been analyzed in Exercise 17.1. If $t_1 + t_2 < 0$, (19.24) holds for any rotation about the 3-axis, and (1923) is violated. For example, a strut subject to pressure on its two ends is always unstable if the criterion (19.22) is taken seriously. According to this definition, then, the Euler column is unstable at all loads.

An instability of this kind has long been recognized but discarded as trivial. Critical loads were calculated by use of particular classes of deformations considered to be reasonable, but no concrete rule was laid down. In recent work BEATTY has given a rational criterion for excluding rotations of this kind. Namely, in the definition of stability he restricts the class of competing deformations to those which preserve equilibrium of moments. He has shown that all the major results of the theory of stability retain their strength when this <u>moment condition</u> is imposed.

<u>Exercise 19.6</u>. Prove that if the loads have no axis of equilibrium (Exercise 15.5), the moment condition excludes all pure rotations in the criterion for stability of the traction boundary-value problem, but if there is an axis, rotations about it, and only those, are allowed.

On the basis of this definition of stability, rigorous estimates of critical loads have been found recently.

LECTURE 20: HYPERELASTIC MATERIALS.

Acoustic Definition.

In Lecture 18 a hyperelastic material was defined as an elastic material whose acoustic tensor $Q(n)$ is symmetric for every direction n in every configuration.

$$Q(n) = Q(n)^T \tag{18.17}$$

In such a material, the Piola-Kirchhoff stress T_κ has a potential, $\sigma(F)$, which is called the stored-energy function:

$$T_\kappa = \mathfrak{h}(F) = \rho_\kappa \partial_F \sigma(F) \ . \tag{18.19}$$

Equivalently, by (11.13),

$$T = \rho \partial_F \sigma(F) F^T \tag{20.1}$$

Problem Area.

In this lecture we shall consider some formal aspects of the special case when a stored-energy function exists. In the next lecture, we shall present some theorems which may justify the assumption that it does.

Frame-indifference.

The condition of material frame-indifference for an elastic material is (11.16), namely,

$$\mathfrak{h}(QF) = Q \ \mathfrak{h}(F) \tag{11.16}$$

for all non-singular F and all orthogonal Q. By (18.20), this condition may be expressed in terms of the stored-energy function σ:

$$\partial_{\underset{\sim}{F}}\sigma(\underset{\sim}{Q}\underset{\sim}{F}) = \underset{\sim}{Q}\partial_{\underset{\sim}{F}}\sigma(\underset{\sim}{F}) . \tag{20.2}$$

Integration yields

$$\sigma(\underset{\sim}{Q}\underset{\sim}{F}) - \sigma(\underset{\sim}{F}) = \phi(\underset{\sim}{Q}) , \tag{20.3}$$

where ϕ is an arbitrary scalar function. Since, however, σ affects the stress only through its gradient, there is no loss in generality in setting $\phi(\underset{\sim}{Q}) = 0$. Then (20.3) becomes

$$\sigma(\underset{\sim}{Q}\underset{\sim}{F}) = \sigma(\underset{\sim}{F}) , \tag{20.4}$$

asserting that σ is a frame-indifferent scalar. Conversely, if (20.4) holds, so does (11.16), granted (18.20). We have shown that <u>a scalar function $\sigma(\underset{\sim}{F})$ may be the stored-energy function of a hyperelastic material if and only if it is frame-indifferent</u>, to within an inessential additive term.[1]

<u>Exercise 20.1</u>. Prove that

$$\sigma(\underset{\sim}{F}) = \sigma(\underset{\sim}{U}) = \bar{\sigma}(\underset{\sim}{C}), \tag{20.5}$$

say, and

$$\underset{\sim}{T} = \rho\underset{\sim}{F}[\partial_{\underset{\sim}{F}}\sigma(\underset{\sim}{F})]^T . \tag{20.6}$$

<u>Exercise 20.2</u>. Prove that (20.1) is frame-indifferent if and only if $\underset{\sim}{T} = \underset{\sim}{T}^T$. Interpret the result as showing that for hyperelastic materials, frame-indifference is equivalent to the balance of moment of momentum, provided linear momentum be balanced.

---

[1]From (20.3) it is clear that

$$\phi(\underset{\sim}{Q}) = \sigma(\underset{\sim}{Q}) - \sigma(\underset{\sim}{1}) .$$

By an appeal to the compactness of the orthogonal group, NOLL <u>proved</u> that $\sigma(\underset{\sim}{Q}) = \sigma(\underset{\sim}{1})$. Thus $\phi(\underset{\sim}{Q})$, the "inessential additive term," is in fact 0.

Stress Working.

The kinetic energy $\mathcal{K}$ of a body $\mathcal{B}$ in the configuration $\underset{\sim}{\chi}$ is defined by

$$\mathcal{K} \equiv \frac{1}{2}\int_{\mathcal{B}_{\underset{\sim}{\chi}}} \rho\dot{\underset{\sim}{x}}^2\, dv = \frac{1}{2}\int_{\mathcal{B}_{\underset{\sim}{\kappa}}} \rho_{\underset{\sim}{\kappa}}\dot{\underset{\sim}{x}}^2 dv \ , \tag{20.7}$$

Its rate of change then is

$$\dot{\mathcal{K}} = \left(\frac{1}{2}\int_{\mathcal{B}_{\kappa}} \rho_{\underset{\sim}{\kappa}}\ \dot{\underset{\sim}{x}}^2 dv\right)^{\cdot} = \int_{\mathcal{B}_{\underset{\sim}{\kappa}}} \rho_{\underset{\sim}{\kappa}}\dot{\underset{\sim}{x}}\cdot\ddot{\underset{\sim}{x}}\, dv = \int_{\mathcal{B}_{\underset{\sim}{\chi}}} \rho\dot{\underset{\sim}{x}}\cdot\ddot{\underset{\sim}{x}}\ dv \ ,$$

$$= \int_{\mathcal{B}_{\chi}} \dot{\underset{\sim}{x}} \cdot [\rho\underset{\sim}{b} + \operatorname{div} \underset{\sim}{T}]dv \tag{20.8}$$

$$= \int_{\mathcal{B}_{\underset{\sim}{\chi}}} \rho\dot{\underset{\sim}{x}} \cdot \underset{\sim}{b} dv + \int_{\partial\mathcal{B}_{\underset{\sim}{\chi}}} \dot{\underset{\sim}{x}} \cdot \underset{\sim}{t} ds - \int_{\mathcal{B}_{\underset{\sim}{\chi}}} \operatorname{tr}(\underset{\sim}{T}\underset{\sim}{D})dv \ ,$$

where we have used Cauchy's first and second laws, (3.16) and (3.18), and Cauchy's lemma (3.11). Accordingly, the increase of kinetic energy may be regarded as the result of three distinct effects: The working of the body force $\underset{\sim}{b}$ in the interior, the working of the traction $\underset{\sim}{t}$ on the boundary, and the working of the stress tensor in the interior, this last being of amount $-\operatorname{tr}(\underset{\sim}{T}\underset{\sim}{D})$ per unit volume. The negative of this last density is called the <u>stress working</u> w:

$$w \equiv \operatorname{tr}(\underset{\sim}{T}\underset{\sim}{D}) = \operatorname{tr}(\underset{\sim}{T}\underset{\sim}{G}^T) \ . \tag{4.17}$$

We have encountered it already in Lecture 4.

In a hyperelastic material, by (20.1) and (2.13),

$$
\begin{aligned}
w &= \rho\,\mathrm{tr}[\partial_{\mathbf{F}}\sigma(\mathbf{F})\ \mathbf{F}^T\mathbf{G}^T), \\
&= \rho\,\mathrm{tr}[\partial_{\mathbf{F}}\sigma(\mathbf{F})\ \mathbf{F}^T(\dot{\mathbf{F}}\mathbf{F}^{-1})^T] , \\
&= \rho\,\mathrm{tr}[\partial_{\mathbf{F}}\sigma(\mathbf{F})\ \dot{\mathbf{F}}^T] \\
&= \rho\dot{\sigma}
\end{aligned}
\tag{20.9}
$$

This result serves to motivate the name "stored-energy function" for σ, since the material time rate $\dot{\sigma}$ of this function is the stress working.

Exercise 20.3. Prove that every elastic fluid is hyperelastic, and show that the stored-energy function is given by

$$
\sigma = -\int \frac{dp}{\rho} \ . \tag{20.10}
$$

The Two Isotropy Groups.

Let $\mathcal{g}$ be the isotropy group of a hyperelastic material with respect to a particular reference configuration. If $\mathbf{H} \in \mathcal{g}$, then by (6.1) and (20.1)

$$
\begin{aligned}
\partial_{\mathbf{F}}\sigma(\mathbf{F})\mathbf{F}^T &= [\partial_{\mathbf{F}}\sigma(\mathbf{F}\mathbf{H})](\mathbf{F}\mathbf{H})^T \\
&= [\partial_{\mathbf{F}}\sigma(\mathbf{F}\mathbf{H})]\ \mathbf{H}^T\mathbf{F}^T ,
\end{aligned}
\tag{20.11}
$$

for arbitrary $\mathbf{F}$. Cancelling $\mathbf{F}^T$ and integrating, we find that

$$
\sigma(\mathbf{F}) = \sigma(\mathbf{F}\mathbf{H}) + \psi(\mathbf{H}) , \tag{20.12}
$$

for all $\mathbf{F}$. Setting $\mathbf{F} = \mathbf{1}$ yields $\psi(\mathbf{H}) = \sigma(\mathbf{1}) - \sigma(\mathbf{H})$. Hence (20.12) becomes

$$
\sigma(\mathbf{F}) = \sigma(\mathbf{F}\mathbf{H}) + \sigma(\mathbf{1}) - \sigma(\mathbf{H}) . \tag{20.13}
$$

Conversely, differentiation of this equation yields (20.11). We have shown that $\mathbf{H}\in\mathcal{g}$ if and only if (20.13) is satisfied for all non-singular $\mathbf{F}$.

We recall that $\mathcal{g}$ is the group of all deformations that cannot be detected by experiments on the stress. We now define $\mathcal{g}_\sigma$ as the group of all unimodular transformations that cannot be detected by experiments on the stored-energy σ . Formally, $\underset{\sim}{H} \varepsilon \mathcal{g}_\sigma$ if and only if

$$\sigma(\underset{\sim}{F}) = \sigma(\underset{\sim}{F}\underset{\sim}{H}) \tag{20.14}$$

for all non-singular $\underset{\sim}{F}$.

If (20.14) is satisfied, we may put $\underset{\sim}{F} = \underset{\sim}{1}$ and obtain $\sigma(\underset{\sim}{1}) = \sigma(\underset{\sim}{H})$. Hence (20.13) is satisfied also. That is,

$$\mathcal{g}_\sigma \subset \mathcal{g} \tag{20.15}$$

Furthermore, by (20.4), $\sigma(\underset{\sim}{Q}) = \sigma(\underset{\sim}{1})$. Therefore if $\underset{\sim}{H}$ is an orthogonal tensor $\underset{\sim}{Q}$, (20.13) reduces to (20.14). That is, the orthogonal subsets of $\mathcal{g}$ and $\mathcal{g}_\sigma$ are identical:

$$\mathcal{g} \cap \mathcal{o} = \mathcal{g}_\sigma \cap \mathcal{o} \quad . \tag{20.16}$$

Now these results hold for every reference configuration. If the material is a solid, there exists some reference configuration for which $\mathcal{g} \subset \mathcal{o}$. By (20.15), $\mathcal{g}_\sigma \subset \mathcal{o}$. Hence $\mathcal{g} \cap \mathcal{o} = \mathcal{g}$ and $\mathcal{g}_\sigma \cap \mathcal{o} = \mathcal{g}_\sigma$. By (20.16), $\mathcal{g} = \mathcal{g}_\sigma$. If the material is a fluid, $\sigma(\underset{\sim}{F}) = f(\det \underset{\sim}{F})$, so (20.14) is satisfied for all $\underset{\sim}{H}$, and again it follows that $\mathcal{g} = \mathcal{g}_\sigma$. We have proved that <u>in a hyperelastic solid or fluid, the two isotropy groups $\mathcal{g}$ and $\mathcal{g}_\sigma$ are the same</u>. In a fluid crystal, generally, $\mathcal{g} \neq \mathcal{g}_\sigma$. That is, there are some deformations which cannot be detected by measurements on the stress but can be detected by measurements of the stored-energy function. This property does not characterize fluid crystals, however, since in some of them, too, $\mathcal{g} = \mathcal{g}_\sigma$.

<u>Minimum of the Stored-Energy Function</u>.

If $\underset{\sim}{H} \varepsilon \mathcal{g}$, (20.13) is satisfied in particular when $\underset{\sim}{F} = \underset{\sim}{H}^{-1}$. Therefore

$$\sigma(\underset{\sim}{H}^{-1}) - \sigma(\underset{\sim}{1}) = \sigma(\underset{\sim}{1}) - \sigma(\underset{\sim}{H}) \quad . \tag{20.17}$$

If $\mathbf{H}$ is orthogonal, then $\sigma(\mathbf{H}) = \sigma(\mathbf{1})$, and this equation reduces to $0 = 0$. If $\sigma(\mathbf{H}) > \sigma(\mathbf{1})$, then $\sigma(\mathbf{H}^{-1}) < \sigma(\mathbf{1})$, and if $\sigma(\mathbf{H}) < \sigma(\mathbf{1})$, then $\sigma(\mathbf{H}^{-1}) > \sigma(\mathbf{1})$. Suppose now the stored-energy function has a minimum in the sense that $\sigma(\mathbf{F}) > \sigma(\mathbf{1})$ when $\mathbf{F}$ is not orthogonal. Then there can be no non-orthogonal $\mathbf{H}$ satisfying (20.17). At a minimum, the gradient of σ vanishes, and hence so does $\mathbf{T}$, by (20.1).

We have proved the following fundamental theorem on hyperelastic solids: If the stored-energy function has a minimum in a certain configuration, the material is a solid, and that configuration is an undistorted natural state.

We have shown that a strict minimum of σ implies the existence of a natural state, which is of course unique to within a rotation. The converse, however, does not hold. Without some further condition, a natural state need not correspond to minimum energy and hence need not be unique.

Coleman & Noll's Inequality.

COLEMAN & NOLL proposed the inequality

$$\sigma(\mathbf{SF}) - \sigma(\mathbf{F}) - \mathrm{tr}[(\mathbf{S} - \mathbf{1})\mathbf{F}(\partial_{\mathbf{F}}\sigma(\mathbf{F}))^T] > 0 \qquad (20.18)$$

for all positive-definite symmetric tensors $\mathbf{S}$ other than $\mathbf{1}$. We call it the the C-N condition. It asserts a type of restricted convexity of the function $\sigma(\mathbf{F})$. Equivalently, by (18.20),

$$\sigma(\mathbf{SF}) - \sigma(\mathbf{F}) - \frac{1}{\rho_\kappa}\,\mathrm{tr}[((\mathbf{SF})^T - \mathbf{F}^T)\,\mathfrak{h}(\mathbf{F})] > 0 \,. \qquad (20.19)$$

If we interchange $\mathbf{SF}$ and $\mathbf{F}$ and add the resulting inequality to (20.19), we conclude that (17.5) holds. Conversely, if (17.5) holds in a convex domain of the space of invertible tensors, we may join $\mathbf{SF}$ and $\mathbf{F}$ by a line, each point of which differs from $\mathbf{F}$ by a pure stretch:

$$\tilde{\mathbf{S}}\mathbf{F} = \mathbf{F} + \lambda(\mathbf{SF} - \mathbf{F}) \,, \quad 0 \leq \lambda \leq 1 \,, \qquad (20.20)$$

where

$$\tilde{\mathbf{S}} = (1 - \lambda)\mathbf{1} + \lambda\mathbf{S} \,, \qquad (20.21)$$

which is positive-definite and symmetric. Then

$$\begin{aligned}
\sigma(\underset{\sim}{S}\underset{\sim}{F}) - \sigma(\underset{\sim}{F}) &- \mathrm{tr}[(\underset{\sim}{S}\underset{\sim}{F} - \underset{\sim}{F})^T \partial_{\underset{\sim}{F}} \sigma(\underset{\sim}{F})] \\
&= \int_0^1 \frac{d}{d\lambda} \sigma(\tilde{S}\underset{\sim}{F}) d\lambda - \mathrm{tr}[(\underset{\sim}{S}\underset{\sim}{F} - \underset{\sim}{F})^T \partial_{\underset{\sim}{F}} \sigma(\underset{\sim}{F})] , \\
&= \int_0^1 \{\mathrm{tr}[(\underset{\sim}{S}\underset{\sim}{F} - \underset{\sim}{F})^T \partial_{\underset{\sim}{F}} (\tilde{S}\underset{\sim}{F}) \\
&\qquad - \mathrm{tr}[(\underset{\sim}{S}\underset{\sim}{F} - \underset{\sim}{F})^T \partial_{\underset{\sim}{F}} \sigma(\underset{\sim}{F})]\} d\lambda , \\
&= \frac{1}{\rho_{\underset{\sim}{\kappa}}} \int_0^1 \mathrm{tr}[(\tilde{S}\underset{\sim}{F} - \underset{\sim}{F})^T [\mathfrak{h}(\tilde{S}\underset{\sim}{F}) - \mathfrak{h}(\underset{\sim}{F})]\} \frac{d\lambda}{\lambda} .
\end{aligned} \tag{20.22}$$

If the GCN condition (17.5) holds in the convex domain, the last integrand is positive. Therefore (20.18) follows. We have proved that in a hyper-elastic material,

$$\text{C-N} \Leftrightarrow \text{GCN}, \tag{20.23}$$

with the understanding that " $\Rightarrow$ " refers to all pairs $\underset{\sim}{F}$, $\underset{\sim}{S}\underset{\sim}{F}$, while " $\Leftarrow$ " refers to pairs which can be connected by a line (20.20) on which GCN holds.

If the material has a natural state, we may take it as reference configuration and conclude from (20.18) that

$$\sigma(\underset{\sim}{S}) > \sigma(\underset{\sim}{1}) \tag{20.24}$$

if $\underset{\sim}{S}$ is positive-definite, symmetric, and not equal to $\underset{\sim}{1}$. Since $\sigma(\underset{\sim}{F}) = \sigma(\underset{\sim}{U})$ by (20.5), we have shown that

$$\text{C-N} \Rightarrow \text{in a natural state, the stored-energy function has a strict minimum.} \tag{20.25}$$

This is a result complementary to the fundamental theorem on solids. Together, they show that

$$\text{C-N} \Rightarrow \text{only solids can have natural states.} \tag{20.26}$$

We must recall, however, that while the fundamental theorem is perfectly general, the C-N condition is a supplemental <u>a priori</u> inequality, the merits of which are still a subject of discussion.

<u>Exercise 20.4</u>. Prove that in an isotropic elastic material the stored-energy function equals a function of the principal stretches:

$$\sigma(\underset{\sim}{F}) = \hat{\sigma}(v_1, v_2, v_3) \ . \tag{20.27}$$

Prove that

$$GCN_0 \Longleftrightarrow \hat{\sigma} \text{ is convex} \tag{20.28}$$

and that

$$\text{C-N} \Longrightarrow \hat{\sigma} \text{ is convex,} \tag{20.29}$$

but the converse is false. For this last, consider the example

$$\rho_{\underset{\sim}{\kappa}}\hat{\sigma}(v_1, v_2, v_3) = K[\tfrac{1}{2}(v_1^2 + v_2^2 + v_3^2) - (v_1 + v_2 + v_3)] \ , \tag{20.30}$$

and use the fact that

$$GCN \Longrightarrow GCN_0 \ \& \ (A_r \geqq 0) \ , \tag{20.31}$$

which follows from (17.19) and (17.24).

LECTURE 21: WORK THEOREMS IN HYPERELASTICITY.

Nonsense about Perpetual Motion.

In most presentations of the infinitesimal theory of elasticity the stored-energy function plays a great part. Commonly, some sort of argument about work done or perpetual motion avoided is brought in to "prove" that a stored-energy function exists. In the foregoing outline of elasticity you will have noticed that no such function was mentioned until toward the end of Lecture 18, and that only in the last lecture did it come to play any great part. My reluctance to employ a strain energy or the arguments associated with it has been misinterpreted as a claim that real materials do not have energy functions or even that energy is not conserved. This is not my view at all. First, I claim nothing about real materials, either in this context or in any other. Second, it is only plain mathematical common sense to defer making an assumption until that assumption serves some purpose. As is illustrated by the foregoing development, for most aspects of elasticity theory, including special problems, it makes no difference whether or not an energy function exists. Third, I have wished to exert my small influence against the diffusion, unfortunately fostered even by some experts in general mechanics, of the muddy verbiage that spouts forth as soon as thermodynamics is mentioned. The formal structure of the infinitesimal theory has been well known for a long time and stands up uninjured by the bad reasoning and confusion of principle various authors, especially paedagogic ones, feel themselves compelled to use in "explaining" it. The finite theory is another matter. Here incorrect thermodynamic arguments can lead to incorrect results.

In the first place, perpetual motion has nothing to do with the matter. According to the infinitesimal theory for isotropic materials, with which critics of every school are perfectly satisfied, and in which no thermodynamic argument is needed to infer the existence of a stored-energy function, perpetual motion occurs, as it does in any theory taking no account of friction. Therefore, a person who does not like perpetual motion in a theory must reject elasticity *in toto*, with or without a stored-energy function. Second, the possibility of extracting work indefinitely, sometimes called a perpetual motion of the second kind, also has nothing to do with the presence or absence of a stored-energy function, as we shall see below.

In the general theory of elasticity, there are various means of motivating the assumption that a stored-energy function exists, or even of proving its existence within a more general framework of ideas. In each case, naturally, a *proved theorem*, not just a lot of physical talk, forms the basis.

Virtual Work of the Traction on the Boundary.

Consider a one-parameter family of deformations:

$$\mathbf{x} = \boldsymbol{\chi}_{\kappa}(\mathbf{X}, \alpha) , \qquad \alpha_1 \leqq \alpha \leqq \alpha_2 . \tag{21.1}$$

Since α need not be the time, such a family is called a *virtual motion*. We shall write

$$\dot{\boldsymbol{\chi}} \equiv \partial_{\alpha}\boldsymbol{\chi}_{\kappa}(\mathbf{X}, \alpha) , \tag{21.2}$$

for the *virtual velocity*. If α is in fact the time, $\dot{\boldsymbol{\chi}} = \dot{\mathbf{x}}$, the velocity. In an elastic body $\mathcal{B}$ a traction $\mathbf{t}_{\kappa}(\mathbf{X}, \alpha)$ on ∂B_{κ} is determined from (21.1) by the constitutive equation (11.15). The *virtual work* done by this traction in the virtual motion is, by definition,

$$W_{12} \equiv \int_{\alpha_1}^{\alpha_2} \left(\int_{\partial B_{\kappa}} \dot{\boldsymbol{\chi}} \cdot \mathbf{t}_{\kappa} \, ds \right) d\alpha . \tag{21.3}$$

Before going a step further we must post a blazing placard: Even if α is the time and the elastic body is made to undergo (21.1) as a real motion,

W_{12} is generally NOT the actual work done

by the forces effecting the motion. Why not? Because (21.1) generally gives rise to an acceleration $\ddot{\mathbf{x}}$ and also by (11.15) to a stress $\mathbf{T}$ in the elastic body, and by Cauchy's first law (3.16) a particular and uniquely determined body force $\underset{\sim}{b}$ will be required in order to make the motion occur. In the unlikely case that the right $\underset{\sim}{b}$ could be produced in the laboratory, it, too, will do work. In any honest attempt to discuss perpetual motion of the second kind, <u>all the work done</u> must be taken into account. Therefore, as stated above, <u>no argument about the virtual work of the traction alone can determine whether or not a perpetual motion of the second kind occurs</u>.

<u>Homogeneous Processes in Homogeneous Bodies</u>.

If the deformation gradient $\underset{\sim}{F}$ as calculated from (21.1) is independent of $\underset{\sim}{X}$, we regard the virtual motion from α_1 to α_2 as a curve $\underset{\sim}{F}(\alpha)$ from the point $\underset{\sim}{F}(\alpha_1)$ to the point $\underset{\sim}{F}(\alpha_2)$ in the space of non-singular tensors $\underset{\sim}{F}$. If $\underset{\sim}{F}(\alpha_1) = \underset{\sim}{F}(\alpha_2)$, the virtual motion is said to be <u>closed</u>. If the elastic body is homogeneous, then $\underset{\sim}{T}_{\kappa}$ is independent of $\underset{\sim}{X}$ also: $\underset{\sim}{T}_{\kappa} = \underset{\sim}{T}_{\kappa}(\alpha)$. Since $\operatorname{Div} \underset{\sim}{T}_{\kappa} = \underset{\sim}{0}$, (21.3) yields

$$
\begin{aligned}
W_{12} &= \int_{\alpha_1}^{\alpha_2} \left(\int_{B_\kappa} \operatorname{Div}(\dot{\underset{\sim}{X}}\, \underset{\sim}{T}_\kappa)\, dv \right) d\alpha \,, \\
& \int_{\alpha_1}^{\alpha_2} \left(\int_{B_\kappa} \operatorname{tr}(\underset{\sim}{T}_\kappa^T \dot{\underset{\sim}{F}})\, dv \right) d\alpha \,, \\
&= V(B_\kappa) \int_{\alpha_1}^{\alpha_2} \operatorname{tr}(\underset{\sim}{T}_\kappa^T \dot{\underset{\sim}{F}})\, d\alpha \,, \\
&= V(B_\kappa) \int_{\underset{\sim}{F}(\alpha_1)}^{\underset{\sim}{F}(\alpha_2)} \operatorname{tr}(T_\kappa^T\, d\underset{\sim}{F}) \,,
\end{aligned}
\tag{21.4}
$$

where $(11.11)_1$ has been used, where $V(B_\kappa)$ is the volume of $\mathcal{B}_\kappa$, and where the last integral is a line integral along the path from $\underset{\sim}{F}(\alpha_1)$ to $\underset{\sim}{F}(\alpha_2)$.

In what follows, we shall assume the reference configuration so chosen that $\det \underset{\sim}{F} > 0$.

The First Work Theorem.

The first work theorem may be read off from (21.4) by using well known facts about line integrals: For a homogeneous elastic body, the following three assertions are equivalent:

1. The virtual work of the traction on the boundary is non-negative in any closed motion.

2. The virtual work of the traction on the boundary is the same in any two homogeneous virtual motions with the same initial and final deformation gradients.

3. The material is hyperelastic, and in a homogeneous virtual motion from $\underset{\sim}{F}(\alpha_1)$ to $\underset{\sim}{F}(\alpha_2)$

$$W_{12} = \mathcal{M}(B)[\sigma(\underset{\sim}{F}(\alpha_2)) - \sigma(\underset{\sim}{F}(\alpha_1))] \ , \tag{21.5}$$

where $\mathcal{M}(\mathcal{B})$ is the mass of $\mathcal{B}$ and σ is the stored-energy function.

We must notice that in a hyperelastic body in an actual motion, the volume integral of $\rho_{\underset{\sim}{\kappa}}\sigma$ does not generally give the total work done. W_{12} may be calculated from $(21.4)_1$, which no longer reduces to $(21.4)_2$. Cf. the related result (20.9).

There is sure to be someone who rises and says, "I don't accept your claim that W_{12} is not the actual work done. In a homogeneous process in a homogeneous body, $\underset{\sim}{T}$ is homogeneous, so $\operatorname{div} \underset{\sim}{T} = \underset{\sim}{0}$, so the body force $\underset{\sim}{b}$ equals the acceleration $\ddot{\underset{\sim}{x}}$, and if the process is slow enough, accelerations can be neglected, so there is no work done beyond W_{12}." Such people cannot be answered. It is best to bow to their wisdoms and let them depart. When they are gone, we can remark to each other that in order for the acceleration to be zero in a homogeneous motion, by (5.6) we see that the only paths allowed in the space of non-singular tensors $\underset{\sim}{F}$ are straight lines: $\underset{\sim}{F}(\alpha) = \underset{\sim}{F}_o(1 + \alpha \underset{\sim}{F}_1)$. Thus if the accelerations are "negligible," in order to apply statement 1 of the theorem we must find a "nearly" closed straight line. To apply statement 2, we must find two different straight line segments that "nearly" connect the same two endpoints, in defiance of Euclid. If we consider two paths which are both "nearly" straight from $\underset{\sim}{F}(\alpha_1)$ to $\underset{\sim}{F}(\alpha_2)$, W_{12} will have "nearly" the same value for each by mere continuity, without any theorem. People who can handle all these "nearlys" have no need to call the theory of finite elastic strain to their aid. They can prove that 2 = 1 without it.

The Second Work Theorem.

According to CAPRIOLI's work theorem, if an elastic body has a homogeneous configuration $\underset{\sim}{\kappa}$ such that the virtual work done in every homogeneous motion from $\underset{\sim}{\kappa}$ is non-negative, then the material is hyperelastic, its stored-energy has a weak minimum at $\underset{\sim}{\kappa}$, and $\underset{\sim}{\kappa}$ is a natural state. To prove this theorem, let the deformation gradient $\underset{\sim}{F}$ be taken with respect to $\underset{\sim}{\kappa}$, so that $\underset{\sim}{F} = \underset{\sim}{1}$ there, and let $\mathcal{C}$ be any closed homogeneous virtual motion. If the point $\underset{\sim}{F} = \underset{\sim}{1}$ does not lie on $\mathcal{C}$, connect it by a path p to some point $\underset{\sim}{F}_o$ which does lie on $\mathcal{C}$. Then $p + \mathcal{C} - p$ is a closed virtual motion from $\underset{\sim}{\kappa}$ back to $\underset{\sim}{\kappa}$. By hypothesis, for this motion $W_{12} \geqq 0$. But W_{12} for this motion is W_{12} for $\mathcal{C}$. Thus alternative 1 in the first work theorem holds. By alternative 3, the material is hyperelastic, and by (21.5) the hypothesis of CAPRIOLI's theorem takes the form

$$\sigma(\underset{\sim}{F}) \geqq \sigma(\underset{\sim}{1}) \ . \tag{21.6}$$

Hence $\partial_{\underset{\sim}{F}}\sigma(\underset{\sim}{F}) = \underset{\sim}{0}$ when $\underset{\sim}{F} = \underset{\sim}{1}$. Therefore $\underset{\sim}{\kappa}$ is a natural state.

The second work theorem gives an energetic criterion sufficient that there be a natural state as well as a stored-energy function. The inequality (21.6) is weaker than (20.24): The minimum has not been proved to be strict, nor can it be, as is shown by the result of the following exercise.

Exercise 21.1. Prove that the stored-energy function

$$\sigma(\underset{\sim}{F}) = K\left(\frac{1}{J} - 1\right)^2 \tag{21.7}$$

does not satisfy (20.24) but does satisfy (21.6) and defines an elastic fluid with a natural state.

The Third Work Theorem.

Let W^*_{12} be the virtual work corresponding to a homogeneous virtual motion $\underset{\sim}{F}(\alpha)$ from $\underset{\sim}{F}(\alpha_1)$ to $\underset{\sim}{F}(\alpha_2)$ when, disregarding the constitutive equation, we hold $\underset{\sim}{T}_{\underset{\sim}{\kappa}}$ at a fixed value $\underset{\sim}{T}^*_{\underset{\sim}{\kappa}}$. By (21.4)

$$W^*_{12} = V(B_{\underset{\sim}{\kappa}})\,\mathrm{tr}\,\{\underset{\sim}{T}^{*T}_{\underset{\sim}{\kappa}}[\underset{\sim}{F}(\alpha_2) - \underset{\sim}{F}(\alpha_1)]\} \quad . \tag{21.8}$$

If $\underset{\sim}{T}^*_{\underset{\sim}{\kappa}} = \underset{\sim}{\mathfrak{h}}(\underset{\sim}{F}(\alpha_1))$, its initial value, W^*_{12} is the virtual work corresponding to dead load (cf. Lecture 19). According to COLEMAN's work theorem, the following two statements are equivalent for an elastic material deformed from a homogeneous configuration:

1. In any motion whose termini differ by a non-identical pure stretch,

$$W_{12} > W^*_{12} \quad . \tag{21.9}$$

2. The material is hyperelastic, and its stored-energy function satisfies the C-N inequality (20.18).

To prove COLEMAN's theorem, we notice first that for a closed motion, $W^*_{12} = 0$ by (21.8), while by (21.9) $W_{12} \geqq W^*_{12}$. Hence in any closed motion $W_{12} \geqq 0$. By the first work theorem, the material is hyperelastic. By (21.5), (21.8), and (18.20), the hypothesis (21.9) assumes the form

$$\sigma(\bar{\underset{\sim}{F}}) - \sigma(\underset{\sim}{F}) > \frac{1}{\rho_{\kappa}}\,\mathrm{tr}\,\{(\bar{\underset{\sim}{F}} - \underset{\sim}{F})\partial_{\underset{\sim}{F}}\sigma(\underset{\sim}{F})^T\} \tag{21.10}$$

if $\bar{\underset{\sim}{F}} = \underset{\sim}{S}\underset{\sim}{F}$, where $\underset{\sim}{S}$ is a positive-definite symmetric tensor other than $\underset{\sim}{1}$. This is COLEMAN & NOLL's inequality (20.19).

COLEMAN's theorem may be interpreted in terms of restricted stability and restricted convexity. The assumption (21.9) would assert even more (in fact impossibly more) than stability if the restriction $\underset{\sim}{F}(\alpha_2) = \underset{\sim}{S}\underset{\sim}{F}(\alpha_1)$ were not laid down. Similarly, without this restriction (20.18) would assert that $\sigma(\underset{\sim}{F})$ is convex. COLEMAN's theorem serves to give a statical motivation for assuming that an elastic material is hyperelastic and satisfies the C-N inequality.

We must not join the "physical" writers on elasticity in confusing the work theorems with thermodynamic arguments. While energy is concerned, heat and temperature are not. It is impossible to prove thermodynamic theorems without first introducing thermodynamics; that is, more quantities and assumptions are needed before any link between existence of a stored-energy function and the "second law of thermodynamics" can be made. A thermodynamic basis for hyperelasticity will be given below in Lecture 28; a thermostatic one, in Lecture 30.

PART IV. FADING MEMORY

LECTURE 22. PRINCIPLES OF FADING MEMORY

Statics and Elasticity.

In Lecture 11 we have seen that the theory of static elasticity and the statics of simple materials are one and the same thing. The stress in an elastic material is given by

$$\underset{\sim}{T} = \underset{\approx\kappa}{\mathfrak{g}}(\underset{\sim}{F}, \underset{\sim}{X}) \;, \tag{11.2}$$

no matter how the present deformation $\underset{\sim}{F}$ from the reference configuration $\underset{\sim}{\kappa}$ has come about in time. In any simple material that has always been at rest, a result of just the same form holds.

In a real situation, we can never be sure that a specimen of material has been at rest forever. On the contrary, we have every reason to suspect that no specimen has been motionless since the beginning of time, and in fact we apply static theory to specimens which we ourselves have moved and perhaps deformed a few days or hours or even seconds ago. The widespread usefulness of static theory gives evidence of another material property: fading memory. The materials of physical experience forget their sufficiently long past deformation and behave _as if_ they had been at rest forever.

It is this property of fading memory in real materials that makes elasticity such a useful theory, despite the fact that it overlooks entirely many obvious material properties such as viscosity and relaxation. This usefulness is not confined to statics. In motion and changing deformation, too, elasticity is often found to give an accurate representation of the behavior of some materials. It is enough that the time required for the particular material to forget the effects of its past be short in comparison with the time scale of the phenomenon. How great these times are depends upon the material, its deformation history, and the phenomenon being studied.

Once we understand these simple ideas, we see fading memory as itself a property of a material that can be given specific mathematical form in terms of the constitutive functional.

Memory of an Elastic Material.

In regard to what we have just said, the elastic materials themselves are rather strange and untypical. A number of statements have been made about the memory of an elastic material in literature, most of which are correct but misleading. A common one is that an elastic material has a perfect memory for its reference configuration and does not remember any other one. This statement is correct in that we can always calculate the stress by knowing the deformation from the reference configuration, but it is misleading in that it does not reflect the arbitrariness of the reference configuration, which may be one occupied by the material a thousand years ago or one occupied one microsecond ago or one never occupied at all. If we choose as reference a configuration occupied at some fixed time t_o, we are justified in saying that the elastic material shows no recollection whatever of any experience it had at any earlier time. In this sense, an elastic material forgets its past so abruptly that it could be described as having not fading memory but failing memory.

Idea of Fading Memory.

It is more typical of a material to have slowly fading memory, not forgetting entirely its past, but rather responding less to long past deformations than to recent ones. In particular, the stress in a material which has been "nearly" at rest for a long time should be "nearly" the static stress.

To make this idea specific, we write the constitutive equation of a simple material in NOLL's form (4.7):

$$\bar{\underset{\sim}{T}} = \underset{\sim}{\mathfrak{K}}(\bar{\underset{\sim}{C}}{}^t_t \; ; \underset{\sim}{C}(t)) , \tag{4.7}$$

where for any tensor $\underset{\sim}{A}$ the notation $\bar{\underset{\sim}{A}}$ is defined by $\bar{\underset{\sim}{A}} = \underset{\sim}{R}(t)^T \underset{\sim}{A}\underset{\sim}{R}(t)$, $\underset{\sim}{R}(t)$ being the present rotation tensor with respect to some fixed reference configuration, and where $\underset{\sim}{C}{}^t_t$ is the history of the relative right Cauchy-Green tensor up to the time t. If $\underset{\sim}{C}{}^t_t = \underset{\sim}{1}$, the material has always been at rest, and

$$\mathfrak{R}(\mathbf{1}\ ;\ \mathbf{C}(t)) = \mathfrak{f}(\mathbf{C}(t))\ , \tag{22.1}$$

where $\mathfrak{f}$ is a function. Set

$$\mathbf{L}^t(s) \equiv \mathbf{C}_t^t(s) - \mathbf{1}\ , \tag{22.2}$$

so that the rest history is $\mathbf{L}^t(s) = \mathbf{0}$, and

$$\mathfrak{H}(\bar{\mathbf{L}}^t;\ \mathbf{C}(t) \equiv \mathfrak{R}(\bar{\mathbf{C}}_t^t;\ \mathbf{C}(t)) - \mathfrak{f}(\mathbf{C}(t))\ . \tag{22.3}$$

Then (4.7) assumes the form

$$\bar{\mathbf{T}} = \mathfrak{f}(\mathbf{C}(t)) + \mathfrak{H}(\bar{\mathbf{L}}^t;\ \mathbf{C}(t)) \tag{22.4}$$

where $\mathfrak{H}$ vanishes at the rest history:

$$\mathfrak{H}(\mathbf{0};\ \mathbf{C}(t)) = \mathbf{0}\ . \tag{22.5}$$

The general constitutive equation in the form (22.3) gives the stress as the sum of two: the "elastic" stress, namely, the stress that would exist if the material had been at rest forever in its present configuration, and a "frictional" or accumulative stress, resulting from the change of deformation it has suffered in the past or is presently suffering.

A rough statement of fading memory is then that the frictional stress is nearly zero if the material is nearly at rest:

$$\mathbf{L}^t(s) \simeq \mathbf{0} \Rightarrow \mathfrak{H}(\bar{\mathbf{L}}^t;\ \mathbf{C}(t)) \approx \mathbf{0}\ . \tag{22.6}$$

In other words, the functional $\mathfrak{H}$ is continuous at the argument function $\bar{\mathbf{L}}^t = \mathbf{0}$. However, this statement is not precise. There are many ways of defining the magnitude of a function, and each of these gives rise to a different concept of continuity. We are interested in the physical interpretation of the results, so we wish to choose a measure of the magnitude of a deformation history $\mathbf{L}^t(s)$ that will reflect the relative unimportance we wish to assign the past.

Obliviating Measure. Recollection.

As usual, we take the magnitude of a tensor $\mathbf{A}$ as given by $|\mathbf{A}| = \sqrt{\mathrm{tr}(\mathbf{A}\ \mathbf{A}^T)}$, and we define the norm $\|\mathbf{A}\|$ of a tensor function

$\underset{\sim}{A}(s)$ of a real variable s by an integral with respect to a Lebesgue-Stieltjes measure μ on the real line:

$$||\underset{\sim}{A}|| \equiv \sqrt{\int_{s=-\infty}^{s=+\infty} |\underset{\sim}{A}(s)|^2 d\mu} \quad . \tag{22.7}$$

The measure, in turn, is generated by a real, non-decreasing function $\sigma(s)$ such that

$$\sigma(s-0) = \sigma(s),$$

$$\mu\{[a,b)\} = \sigma(b) - \sigma(a) \; . \tag{22.8}$$

All this is general. We wish to apply (22.7) to histories, which are functions of s defined only at present and past times, namely when $s \geqq 0$, and we wish the total past to contribute only a finite amount to the magnitude we assign to a bounded history. Accordingly, we shall call the measure μ an <u>obliviating measure</u> if

$$\begin{aligned} &\sigma(s) \equiv 0 \quad \text{when} \quad s \leqq 0 \; , \\ &\lim_{s\to\infty} \sigma(s) = M < \infty \; . \end{aligned} \tag{22.9}$$

From (22.9) it follows that translation of any interval infinitely far back into the past decreases its measure to 0:

$$\lim_{\alpha\to o} \mu\{[a + \alpha, b + \alpha]\} = 0 \; . \tag{22.10}$$

The norm (22.7) if calculated with respect to an obliviating measure is called the <u>recollection</u> of the history $\underset{\sim}{A}(s)$

Let us now find conditions such as to make a deformation history $\bar{\underset{\sim}{L}}^t$ have small recollection. Since $|\bar{\underset{\sim}{L}}^t| = |\underset{\sim}{L}^t|$, it follows that $\|\underset{\sim}{L}^t\| = \|\bar{\underset{\sim}{L}}^t\|$ so it is all one whether we speak of $\underset{\sim}{L}^t$ or $\bar{\underset{\sim}{L}}^t$ in this regard. Since

$$\|\underset{\sim}{L}^t\| = \sqrt{\int_{s=o}^{s=\infty} |\underset{\sim}{L}^t(s)|^2 d\mu} \; , \tag{22.11}$$

clearly $\|\mathbf{L}^t\|$ will be a small number if $\|\mathbf{L}^t(s)\|$ is small for all s. By (22.2), then, a deformation history that has deviated little from rest throughout the entire past will have small recollection, as was desired. Second, since the obliviating measure gives little weight to the part of the s-axis where s is large, a deformation history such that $|\mathbf{L}^t(s)|$ is small for the recent past, though perhaps large in the distant past, has small recollection. This corresponds to the case when the material has been nearly at rest for a long time, although it may have a violent ancient history. Finally, if $|\mathbf{L}^t(s)|$ is large only on a certain set of small measure, as for example a small interval in which $\sigma(s)$ is continuous, the recollection of $\mathbf{L}^t$ will be small. This case corresponds to a large deformation during a very brief period.

We have described three kinds of deformation histories that have small recollection. Since these are kinds of deformations we expect to have little effect on the present behavior of many real materials, we are encouraged to use the recollection (22.11) to define fading memory as an assertion of continuity.

From now on we shall consider only those histories $\mathbf{L}^t(s)$ with finite recollections. These include all histories of bounded magnitude, and certain unbounded histories as well. We shall suppose laid down a certain obliviating measure, which is kept fixed. This measure defines the recollection, and the recollection defines a topology in the space of histories of bounded recollection. Having a topology in this vector space, we can define continuity and differentiability at the origin. Since the origin corresponds to the rest history, we shall be making assumptions and proving theorems concerning materials nearly at rest.

<u>Weak Principle of Fading Memory</u>.

COLEMAN & NOLL's <u>weak principle of fading memory</u>, as generalized by WANG, is as follows: <u>The functional $\mathfrak{H}$ in the form</u> (22.4) <u>of the general constitutive equation of a simple material is continuous at the rest history</u>. That is, by (22.5),

$$\mathfrak{H}(\bar{\mathbf{L}}^t;\ \mathbf{C}(t)) = o(\|\mathbf{L}^t\|). \tag{22.12}$$

In a material that satisfies this principle, for any sequence of histories $\underset{\sim}{L}^t(s)$ which approach the rest history $\underset{\sim}{0}$ in the sense of recollection, the stress in the material approaches the static stress.

The principle of fading memory is, of course, a definition. Some familiar materials satisfy it; others do not. For example, the elastic material satisfies the principle trivially, since $\underset{\sim}{\mathfrak{H}} \equiv \underset{\sim}{0}$. For a second example, consider the Navier-Stokes fluid, or, more generally, any material in which the stress is determined by $\underset{\sim}{F}, \dot{\underset{\sim}{F}}, \ddot{\underset{\sim}{F}}, \ldots, \overset{(n)}{\underset{\sim}{F}}$. Since the derivatives $\dot{\underset{\sim}{F}}, \ddot{\underset{\sim}{F}}, \ldots$ may be of any magnitude in a history with arbitrarily small recollection, (22.12) is not satisfied, in general, by these materials. Nevertheless, the Navier-Stokes fluid has fading memory in the intuitive sense that long past events have no effect on the present stress; indeed, the stress is unaffected by any deformation at any finitely past time. This example shows that the COLEMAN-NOLL definition does not include every possible concept of fading memory. It remainds us that their principle is not a general law of mechanics but a definition of a class of materials, a class which has been found useful because of the number of important theorems proved about it.

Stress Relaxation.

On the basis of the concept of fading memory alone we can prove a fundamental theorem, the theorem on stress relaxation. This theorem asserts that if a deformation occurs far enough in the past it has no or little effect on the present stress. In the physical context to which we are accustomed, we hold the strain fixed and let time go on. Mathematically the idea is equivalent to looking at the present state of deformation and pushing the process of arriving at this state back into the past.

Let the history of a deformation process be described by $\underset{\sim}{L}^t(s)$, and let the static continuation $\underset{\sim}{L}_n(s)$ of this history be defined by

$$\underset{\sim}{L}_n(s) \equiv \begin{cases} \underset{\sim}{0} & \text{if} \quad 0 \leq s \leq n\,, \\ \underset{\sim}{L}^t(s-n) & \text{if} \quad s > n \end{cases} \tag{22.13}$$

The stress-relaxation theorem states that if we take any deformation history $\underset{\sim}{L}^t(s)$ of finite recollection and find the stress $\underset{\sim}{T}_n$ corresponding to its static continuation $\underset{\sim}{L}_n(s)$, then as n approaches to infinity, $\underset{\sim}{T}_n$ approaches the static stress $\underset{\sim}{T}_o$:

$$\lim_{n\to\infty} \underset{\sim}{T}_n = \underset{\sim}{T}_o \ . \tag{22.14}$$

This fundamental theorem requires some additional assumption, either on the obliviating measure or on the histories considered. COLEMAN & NOLL's assumption, as generalized by WANG, is that the measure of any Borel set is not to increase by a shift to the right. That is, if T is a Borel set on the real line, and

$$T_\alpha = \{s|s-\alpha \ \ \varepsilon T\}, \tag{22.15}$$

then

$$\mu(T_\alpha) \leqq \mu(T) \quad \text{for all positive } \alpha \ . \tag{22.16}$$

We can easily construct obliviating measures that do satisfy this condition and others that do not. An example which satisfies it is the measure defined by a function $\sigma(s)$ which is differentiable and has a monotone derivative. Note that an assumption like this is in fact an assumption about the material, because the principle of fading memory, itself an assumption, requires the constitutive functional to have certain properties stated in terms of an assigned measure.

Now I shall outline the proof of the stress-relaxation theorem, leaving the details as an exercise.

Exercise 22.1. (i) Prove that if $\underset{\sim}{L}^t(s)$ is of finite recollection, so is its static continuation $\underset{\sim}{L}_n(s)$. Note also that

$$\lim_{n\to\infty} \mu\{s \mid \ |\underset{\sim}{L}_n(s)| \neq 0\} = 0 \ . \tag{22.17}$$

(ii) Because of (1) and the assumption (22.16)

$$||\underset{\sim}{L}_n(s)|| \to 0 \quad \text{as} \quad n \to \infty \tag{22.18}$$

(iii) Complete the proof of the stress-relaxation theorem.

In order to prove this theorem for arbitrary deformation histories measure of a set of times is not to increase by a translation into the past. We can weaken the requirements on the obliviating measure if we strengthen the requirements on the deformation history. If we consider only somewhat better behaved $\underset{\sim}{L}^t$, we can prove the theorem for a general obliviating measure, without the restriction (22.16).

<u>Exercise 22.2</u>. By the use of Lebesgue's dominated convergence theorem, prove the stress-relaxation theorem without the assumption of (22.16), if any of the following conditions holds.

(i) $\underset{\sim}{L}^t(s)$ is essentially bounded, <u>i.e.</u>, it is bounded except on a set of measure zero.

(ii) $\underset{\sim}{L}^t(s)$ is a non-decreasing function of s.

(iii) $\underset{\sim}{L}^t(s)$ is essentially bounded if $s \leqq m$, where m is some positive number, and $\underset{\sim}{L}^t(s)$ is a non-decreasing function of s if $s > m$.

Consider now an important special case, the case of a strain impulse, where the history is

$$\underset{\sim}{F}(t) = \begin{cases} \underset{\sim}{F} & \text{when} \quad t > t_o \\ \underset{\sim}{1} & \text{when} \quad t < t_o \,. \end{cases} \tag{22.19}$$

In this history the material is held in one fixed configuration until a certain time t_o, when it is suddenly deformed into another configuration and held there forever. From the definition (3.26) of a simple material we wee at once that the stress $\underset{\sim}{T}$ is given by

$$\underset{\sim}{T} = \underset{\sim}{\mathfrak{G}}(\underset{\sim}{F}^t) = \underset{\sim}{\mathcal{G}}(\underset{\sim}{F}, t - t_o) \quad \text{if} \quad t > t_o \,, \tag{22.20}$$

where $\underset{\sim}{\mathcal{G}}$ is now a function instead of a functional because the whole deformation history is specified by two things: the amount $\underset{\sim}{F}$ of the strain impulse and the time t_o at which it occurred. The result (22.20) says then that in a strain impulse any simple material behaves like an elastic material except that the response function depends on time. So far this has nothing to do with fading memory. If, in addition, the material has fading memory, it follows from the stress-relaxation theorem that

$$\underset{\sim}{\mathcal{G}}(\underset{\sim}{F}, t - t_o) \rightarrow \underset{\sim}{\mathcal{G}}(\underset{\sim}{F}) \quad \text{as} \quad t_o \rightarrow \infty \,. \tag{22.21}$$

That is, if the strain impulse occurred long enough in the past, the stress must be close to the static stress. This fact, first observed by RIVLIN through a more involved process of reasoning, we see here as resulting from the simplest application of the theory of fading memory in COLEMAN & NOLL's sense.

LECTURE 23: POSITIONS OF THE CLASSICAL THEORIES OF CONTINUA

Higher-Order Fading Memory.

The results obtained in the last lecture show that we are on the right track. The stress-relaxation theorem is something we expect to get under certain restrictions on the material or on the deformation history, or on both. If we did not get this from our formulation of fading memory, we should not be looking at the right thing. With this much reassurance, we can face the question of how to calculate a second approximation to the constitutive equation if we are not satisfied with (22.12), which in effect gives the elastic stress only. Such an approximation is gotten by what amounts to the use of Taylor-series expansion of the constitutive functional in the neighborhood of the rest history. However, Taylor's theorem concerns a function, and we are dealing here with functionals. The formal theory is somewhat complicated. I shall outline the results without going into details.

We write the constitutive equation of a simple material again in the form

$$\tilde{\mathbf{T}} = \mathfrak{t}\,(\mathbf{C}(t)) + \mathfrak{H}(\bar{\mathbf{L}}^t;\ \mathbf{C}(t)) \ . \tag{22.4}$$

The nth-order principle of fading memory is the assumption that at $\bar{\mathbf{L}}^t = \mathbf{0}$, $\mathfrak{H}$ is n-times Fréchet differentiable. Fréchet differentiation of functionals is a generalization of differentiation of functions, and to assume a functional Fréchet differentiable to the nth order is the same as to assume it has a kind of n-term Taylor expansion with remainder.

Specifically, in the present case,

$$\mathfrak{H}(\bar{\mathbf{L}}^t;\ \mathbf{C}(t)) = \sum_{i=1}^{n} \mathfrak{H}_i\,(\bar{\mathbf{L}}^t;\ \mathbf{C}(t)) + o(||\mathbf{L}^t||^n) \ , \tag{23.1}$$

where $\underset{\sim}{\mathfrak{H}}_i$ is a bounded homogeneous polynomial functional of $\bar{\underset{\sim}{L}}^t$ of degree i, depending on $\underset{\sim}{C}(t)$ as a parameter.

Like the Taylor expansion, which expresses $f(x)$ as a sum of homogeneous polynomial functions x^i, of degrees 0, 1, ... , n, with an error which approaches 0 faster than x^n, the Fréchet expansion replaces the given functional $\underset{\sim}{\mathfrak{H}}(\bar{\underset{\sim}{L}}^t)$ by a sum of simpler functionals of $\bar{\underset{\sim}{L}}^t$, with an error which approaches $\underset{\sim}{0}$ faster than the nth power of the recollection $||\underset{\sim}{L}^t||$. The Taylor expansion, in the form in which we have just stated it, gives no information about $f(x)$ at a single point x; rather, it tells us that if we choose any sequence x_i such that $x_i \to 0$, then the error made by the approximate formula approaches 0 faster than $|x_i|^n$. Similarly, the Fréchet expansion (23.1) tells us nothing at all about the value of $\underset{\sim}{\mathfrak{H}}$ for a particular deformation history $\underset{\sim}{L}^t$. Rather, it tells that if we construct a sequence or <u>family</u> of deformation histories whose recollection approaches 0 , then, ultimately, the error made in replacing $\underset{\sim}{\mathfrak{H}}$ by a certain sum of polynomial functionals of degrees up to n approaches $\underset{\sim}{0}$ faster than $||\underset{\sim}{L}^t||^n$.

Thus in all applications of the assumption that a material obeys the nth-order principle, we shall have to construct families of $\underset{\sim}{L}^t$ such that $||\underset{\sim}{L}^t|| \to 0$. This necessity was illustrated in the last lecture in the proof of the stress-relaxation theorem. We may interpret that theorem as providing circumstances in which the general constitutive equation may be replaced, approximately, by that of an elastic material. The theorem asserts that in the family $\underset{\sim}{L}_n(s)$ defined from a given $\underset{\sim}{L}^t(s)$ by (22.13), <u>viz</u>,

$$\underset{\sim}{L}_n(s) \equiv \begin{cases} \underset{\sim}{0} & \text{if} \quad 0 \leqq s \leqq n \\ \underset{\sim}{L}^t(s) & \text{if} \quad s > n , \end{cases} \tag{22.13}$$

the stress becomes approximately the elastic stress if n is large enough. How large is "large enough" depends on two things: the original deformation history $\underset{\sim}{L}^t(s)$, and the particular material defined by the response functional $\underset{\sim}{\mathfrak{H}}$. To make this fact obvious, consider first the case when $\underset{\sim}{L}^t(s)$ = const., and, second, the case when $\underset{\sim}{\mathfrak{H}} \equiv \underset{\sim}{0}$. In both these cases, the error is 0 for non-negative n. Similarly, the property of higher-order fading memory tells us nothing about the effect of a single motion on a single material. Rather, it tells us how to construct simpler constitutive equations that will be valid asymptotically for certain classes of materials in certain families of deformation histories.

Finite Linear Visco-elasticity.

If a material is assumed to have fading memory of order 1, (23.1) approximates the frictional stress by a bounded homogeneous linear functional. Now the collection of all deformation histories of finite recollection (22.11) is a Hilbert space, and according to the Fréchet-Riesz theorem, every bounded linear functional in a Hilbert space has a representation as an inner product. In the present application, this theorem asserts the existence of a kernel $K(\mathbf{C}, s)$ such that (22.4) and (23.1) assume the form

$$\bar{\mathbf{T}} = \mathbf{f}(\mathbf{C}) + \int_{s=0}^{s=\infty} K(\mathbf{C},s)[\bar{\mathbf{L}}^t(s)]d\mu + o(||\mathbf{L}^t||) . \tag{23.2}$$

The kernel K is a fourth-order tensor function of $\mathbf{C}$ and s. Moreover,

$$\int_{s=0}^{s=\infty} |K(\mathbf{C},s)|^2 d\mu < \infty . \tag{23.3}$$

If we simply omit the error term in (23.2), we obtain a simple class of constitutive equations defining finite linearly visco-elastic materials. The term "finite" refers to the fact that the constitutive equations under consideration are frame-indifferent and hence have meaning for arbitrary deformation histories. As we shall see shortly, a special theory of this type, but acceptable only for infinitesimal deformations, was proposed by BOLTZMANN. Recently it has been the subject of a revival of interest. Only a few years ago, no one knew how to state this theory in a frame-indifferent way. The procedure followed here delivers frame-indifferent approximations effortlessly, so that now it is hard to see how there ever was a problem. The theory of finite linear visco-elasticity was first formulated by COLEMAN & NOLL. It now has a considerable literature, and experimental work applying it to particular materials has been and is being done.

The word "linear" refers to the dependence of the stress upon the past history $\underset{\sim}{L}^t$ of the relative deformation. The nature of the memory of the material is linear in that the frictional stress corresponding to the sum of two different deformation histories leading to the same present deformation from a fixed reference is the sum of the frictional stresses corresponding to each alone. The stress may depend in an arbitrary way on the present deformation $\underset{\sim}{C}(t)$.

Slow Motion and Rivlin-Ericksen Materials.

One way of constructing a family of deformation histories with small recollection is to introduce the concept of slow motion. There is a large and vague literature on slow motion in hydrodynamics, an old favorite for obscure guessing. The first precise concept of slow motion is due to COLEMAN & NOLL. Given a deformation history $\underset{\sim}{L}^t(s)$, define its retardation $\underset{\sim}{L}^\alpha(s)$ as

$$\underset{\sim}{L}^\alpha(s) \equiv \underset{\sim}{L}^t(\alpha s) \ , \quad 0 < \alpha < 1 \ . \tag{23.4}$$

In multiplying all time lapses uniformly by a factor less than 1 we are stretching out the past. From a given history $\underset{\sim}{L}^t(s)$, we construct a family of histories by running over the same sequence of configurations at uniformly slower and slower rates. Intuitively we see that as the retardation factor α approaches zero, the recollection of the retarded history will also approach zero provided the history is smooth. We need an estimate of the rapidity of that approach.

If $\underset{\sim}{L}^t(s)$ has n continuous derivatives at $s = 0$, by Taylor's theorem

$$\underset{\sim}{L}^\alpha(s) = \sum_{j=0}^{n} \frac{s^j \alpha^j}{j!} \underset{\sim}{L}^{(j)}(\tau)\Big|_{\tau=0} + \underset{\sim}{R}_n \ , \tag{23.5}$$

where

$$|\underset{\sim}{R}_n| = o(\alpha^n s^n) \quad \text{as} \quad \alpha s \to 0 \ . \tag{23.6}$$

Exercise 23.1. Let $\underset{\sim}{A}_k$ and $\underset{\sim}{A}_k^\alpha$ be the kth Rivlin-Ericksen tensors of the motion and the retarded motion, respectively, at t. Show that

$$\bar{\underset{\sim}{A}}_k^\alpha = \alpha^k \bar{\underset{\sim}{A}}_k , \quad k = 1, 2, 3, \ldots , \tag{23.7}$$

and hence that

$$\bar{\underset{\sim}{L}}^\alpha(s) = \sum_{j=1}^{n} \frac{s^j}{j!} (-1)^j \bar{\underset{\sim}{A}}_j^\alpha + \bar{\underset{\sim}{R}}_n , \tag{23.8}$$

where again the remainder $\underset{\sim}{R}_n$ satisfies (23.6).

The retardation theorem of COLEMAN & NOLL states that if the original history $\underset{\sim}{L}^t(s)$ is of finite recollection, then for a certain major class of obliviating measures the remainder $\underset{\sim}{R}_n$ in (23.8) has arbitrarily small recollection as $\alpha \to 0$. Specifically,

$$\| \underset{\sim}{R}_n \| = o(\alpha^n) \quad \text{as} \quad \alpha \to 0 . \tag{23.9}$$

While the magnitude of the error $\underset{\sim}{R}_n$ according to (23.6) may grow larger and larger for fixed α as $s \to \infty$, that is, in the distant past, the obliviating measure overcomes the effect of any such growth on the recollection and yields the same order for $\| \underset{\sim}{L}^\alpha \|$ as would follow if the estimate (23.6) were uniform in α. This is far from obvious. While the result (23.9) is simple to state, it is difficult to prove, one of the two or three most difficult theorems in all of continuum mechanics. I have been hoping that some bright person would come along and give a proof of the theorem which is as simple as its statement.

The difficult but powerful theorem of COLEMAN & NOLL enables us to approximate the constitutive equation of a simple material in a striking form. Namely, if we put (23.8) and (23.9) into (23.1) and (22.4), by a straightforward if tedious calculation we can prove that the stress $\underset{\sim}{T}^\alpha$ corresponding to the retarded history $\underset{\sim}{L}^\alpha(s)$ is given by

$$\bar{\underset{\sim}{T}}^\alpha = \underset{\sim}{f}(\underset{\sim}{C}) + \Sigma \underset{\sim}{\ell}_{j_1 \ldots j_k}(\underset{\sim}{C})[\bar{\underset{\sim}{A}}_{j_1}^\alpha, \ldots, \bar{\underset{\sim}{A}}_{j_k}^\alpha] + o(\alpha^n) , \tag{23.10}$$

where the sum runs over all indices $j_1, \ldots, j_k$ such that

$$\begin{aligned} 1 \leq j_1 \leq \cdots \leq j_k \leq n , \\ j_1 + j_2 + \ldots + j_k = n , \end{aligned} \tag{23.11}$$

and where $\underset{\sim}{\ell}_{j_1 \cdots j_k}$ is a multilinear function of the k arguments $\underset{\sim}{\bar{A}}{}^{\alpha}_{j_1}, \ldots, \underset{\sim}{\bar{A}}{}^{\alpha}_{j_k}$.

At first sight (23.10) looks like an impossible result. We began by approximating the frictional stress by a bounded homogeneous linear functional. Now the value of a functional depends upon an argument which is a function, the entire deformation history from $s = 0$ to $s = \infty$. However, (23.10) says that to determine the value of the functional in this case, you do not need to know anything about the history except for an arbitrarily small interval before the present time, since the Rivlin-Ericksen tensors are gotten by time differentiation at the present instant and the multilinear operators $\underset{\sim}{\ell}_{j_1 \cdots j_k}$ depends only on the present deformation. Thus, somehow, the long-range memory back over infinite time is replaced approximately by taking account of enough derivatives at the present instant. We know that logically this cannot be so. The entire past cannot be equated to any number of derivatives at the present. This fact is reflected in the error term; while it is written in (23.10) as $o(\alpha^n)$, we must not forget that it is a functional of the deformation history. The theorem expressed by (23.10) is asymptotic. It is a theorem, not about a single motion, but about a family of motions. It says that, given any motion, by retarding it sufficiently one can justify approximating the constitutive functional of the retarded motion by multilinear function of the time derivatives of the given motion at the present instant. What the retardation process does is to push any severe deformation more and more into the past, which the material does not remember well, and to give the smooth history near the present more and more weight.

According to the remarkable theorem (23.10), the general constitutive equation of a material with long-range fading memory is approximated, in sufficiently retarded motion, by that of a particular material with short-range memory. A material in which $\underset{\sim}{\bar{T}}$ is a function of $\underset{\sim}{C}$ and $\underset{\sim}{\bar{A}}_1, \underset{\sim}{\bar{A}}_2, \ldots, \underset{\sim}{\bar{A}}_n$ is called <u>Rivlin-Ericksen material of complexity n</u>. The special case when $n = 3$ has been encountered in Lecture 7, Eq. (7.29). With the additional requirements expressed by (23.10) and (23.11), the Rivlin-Ericksen material is said to be of <u>grade</u> n. Rivlin-Ericksen materials do not satisfy the principle of fading memory in COLEMAN & NOLL's sense, but in slow motions they serve as approximations to all materials that do.

The case when $n = 0$ is of special interest:

$$\bar{T}^{\alpha} = \mathfrak{f}(C) + o(1) . \tag{23.12}$$

That is, in a sufficiently retarded motion the stresses in any material with fading memory is approximately that of an elastic material. While similar in form to the stress-relaxation theorem, this theorem refers to a different family of histories: slow motions rather than static continuations. For isotropic materials and for fluids, respectively, (23.12) yields

$$\begin{aligned} T^{\alpha} &= \mathfrak{f}(B) + o(1) , \\ T^{\alpha} &= - p(\rho)1 + o(1) , \end{aligned} \tag{23.13}$$

as is immediate from the general theory presented in Lecture 6.

If $n = 1$, (23.10) yields

$$\bar{T}^{\alpha} = \mathfrak{f}(C) + V(C)[\bar{A}_1^{\alpha}] + o(\alpha) , \tag{23.14}$$

where V is a linear operator called the <u>viscosity tensor</u>. This approximate constitutive equation defines the Rivlin-Ericksen material of grade 1, also called the <u>linearly viscous material</u>.

<u>Exercise 23.2</u>. Prove that isotropic and fluid Rivlin-Ericksen materials of complexity n have constitutive equations of the forms

$$T = \mathfrak{f}(B) + \mathfrak{g}(A_1, A_2, \ldots , A_n, B) , \tag{23.15}$$

and

$$T = - p(\rho)1 + \mathfrak{g}(A_1, A_2, \ldots , A_n, \rho) , \tag{23.16}$$

respectively, where the functions $\mathfrak{g}$ are isotropic. In particular, the Rivlin-Ericksen fluid of complexity 1 (or grade 1) is the Navier-Stokes fluid:

$$T = - p(\rho)1 + \lambda(\operatorname{tr} D)1 + 2\mu D . \tag{23.17}$$

Consequently in slow motions of isotropic materials and fluids, respectively,

$$T^{\alpha} = \mathfrak{f}(B) + V(B)[A_1^{\alpha}] + o(\alpha) ,$$

$$T^{\alpha} = - p(\rho)1 + \lambda(\mathrm{tr}D^{\alpha})1 + 2\mu D^{\alpha} + o(\alpha) .$$

Infinitesimal Deformations.

A different application of the theory of fading memory may be made by constructing a family of deformations that have been small throughout the entire past. As in Lecture 14, we set

$$H \equiv \nabla u = F - 1 , \tag{14.2}$$

$$\tilde{E} \equiv \frac{1}{2}(H + H^T), \ \tilde{R} \equiv \frac{1}{2}(H - H^T) , \tag{14.7}$$

and we set

$$\varepsilon = \sup_{s \geq o} |H^t(s)| . \tag{23.19}$$

Thus ε is an upper bound for the magnitude of the deformation gradient throughout the entire past. We shall consider now a family of deformation histories such that $\varepsilon \to 0$.

Exercise 23.3. Prove that

$$L^t(s) = 2[\tilde{E}(t-s) - \tilde{E}(t)] + O(\varepsilon^2) = O(\varepsilon) . \tag{23.20}$$

Thus, as expected, $||L^t|| = O(\varepsilon)$ as $\varepsilon \to 0$, so we may substitute (23.20) into (23.2) and obtain

$$\bar{T} = \mathfrak{k}(C) + 2\int_{s=0}^{s=\infty} K(C,s)[\tilde{E}(t-s) - \tilde{E}(s)]d\mu + o(\varepsilon) . \tag{23.21}$$

This is not all, however, since the kernel $K(C,s)$ may be simplified for infinitesimal deformations, and also $\bar{T}$ can be expressed in terms of T and R.

Exercise 23.4. Prove that

$$\underset{\sim}{C}(t) = \underset{\sim}{1} + 2\underset{\sim}{\tilde{E}}(t) + O(\varepsilon^2) = \underset{\sim}{1} + O(\varepsilon) ,$$

$$\underset{\sim}{R}(t) = \underset{\sim}{1} + \underset{\sim}{\tilde{R}}(t) + O(\varepsilon^2) = 1 + O(\varepsilon) , \tag{23.20}$$

and hence (23.21) assumes the form

$$\underset{\sim}{T}(t) - \underset{\sim}{T}_o = \underset{\sim}{\tilde{R}}(t)\, \underset{\sim}{T}_o - \underset{\sim}{T}_o \underset{\sim}{\tilde{R}}(t)$$

$$+ (L + M(0))[\underset{\sim}{\tilde{E}}(t)]$$

$$+ \int_{s=o}^{s=\infty} \dot{M}(s)[\tilde{E}(t-s)]ds + o(\varepsilon) ,$$

where

$$\underset{\sim}{T}_o = \underset{\sim}{\mathfrak{f}}(\underset{\sim}{1})$$

$$M(s) = -2 \int_{\sigma=s}^{\sigma=\infty} K(\underset{\sim}{1},\sigma)d\mu . \tag{23.24}$$

$\underset{\sim}{T}_o$ is the stress in the reference configuration, and $M(s)$ is the stress-relaxation function. When $M(s) = 0$, (23.23) reduces, apart from the error term to (14.12), the constitutive equation for infinitesimal elastic deformation from a stressed configuration. The terms proportional to $\underset{\sim}{\tilde{R}}$ are the famous convective terms first found by CAUCHY, and L is the linear elasticity. When $\underset{\sim}{T}_o = \underset{\sim}{0}$, the reference configuration is a natural state, and (23.23), again apart from the error term, reduces to the constitutive equation of BOLTZMANN's theory of infinitesimal viscoelasticity.

Positions of the Classical Theories.

We have established four different positions for various classical theories of materials with respect to NOLL's general theory of simple materials.

1. As <u>special materials</u>. The elastic material and the linearly viscous material afford special cases, examples of possible behavior of a material for <u>all</u> deformation histories. The finitely visco-elastic material is another such case. Eulerian fluids are included as a special case in all three of these examples. BOLTZMANN's theory and the infinitesimal theory of elasticity do <u>not</u> fall in line here, since they are not frame-indifferent theories for large deformations.

2. As theories valid for <u>special deformations</u>. Namely, in static deformation all simple materials behave like elastic materials.

3. In <u>slow motions</u> the constitutive equation of a simple material with fading memory of order n is approximated by that of a RIVLIN-ERICKSEN material of grade n. In particular, the elastic material emerges when $n = 0$, the linearly viscous material when $n = 1$. Thus, for example, the NAVIER-STOKES formula is the common first approximation to the constitutive equations of all fluids in sufficiently retarded motion.

4. In <u>infinitesimal deformations</u>. The BOLTZMANN theory, adjusted to allow for initial stress, is the common first approximation to the constitutive equations of all materials with fading memory of order 1 subjected to deformations that are and always have been sufficiently small.

The results according to the third and fourth cases would generally contradict one another for one particular material. According to the former, the stress is determined by a sufficient number of derivatives of the deformation at the present instant. According to the latter, to determine the present stress it is necessary to integrate over the entire past. In fact, both results are approximate, and the circumstance to which they refer are different. In general, theories of the <u>rate type</u> are appropriate to <u>slow motions</u>, while theories of the <u>integral type</u> are appropriate to <u>small deformations</u>. It is advisable to warn the student that no converse has been proved, nor can one be. Theories of the rate type <u>may</u> apply to rapid deformations, as, it seems, the Navier-Stokes equations do for common fluids, and theories of the integral type <u>may</u> apply to large deformations. The importance of COLEMAN & NOLL's theory lies in its positive character. It shows that the classical constitutive equations for continua <u>must</u> hold in certain common cases as approximations to much more general material response.

The range of validity of the approximate formulae is left vague in each case. We know only that by choosing the retardation factor α small enough in the former case, and the bound ε small enough in the latter, we shall get arbitrarily close approximation. How small is "small enough" depends on the material and the deformation history taken as point of departure. Rather than trying to estimate the range of validity of an approximation, it is sometimes easier to improve it. The methods given above can easily be applied to get higher approximations, corrections to the classical constitutive equations. We shall consider one such example and its consequences in the next lecture.

LECTURE 24: THE INCOMPRESSIBLE FLUID OF SECOND GRADE.

Constitutive Equation

By (23.16) and (23.11), the constitutive equation of a fluid of second grade is of the form

$$\mathbf{T} + p\mathbf{1} = \mathfrak{f}_2(\mathbf{A}_1) + \mathfrak{f}_1(\mathbf{A}_2) , \tag{24.1}$$

where $\mathfrak{f}_2$ and $\mathfrak{f}_1$ are isotropic functions of degrees 2 and 1, respectively. By (11.27), the functions $\mathfrak{f}_2$ and $\mathfrak{f}_1$ can be given explicit representations. We shall consider only incompressible fluids, for which p is arbitrary, so that all terms proportional to $\mathbf{1}$ may be absorbed into the term $p\mathbf{1}$. The resulting constitutive equation for the <u>incompressible fluid of second grade</u> is

$$\mathbf{T} = -p\mathbf{1} + \mu\mathbf{A}_1 + \alpha_1\mathbf{A}_2 + \alpha_2\mathbf{A}_1^2 , \tag{24.2}$$

where μ, α_1, and α_2 are constants. According to the general theory outlined in the last lecture, in the limit of retardation of any deformation history of finite recollection, for any fluid with fading memory of second order, (24.2) approximates the stress better than does the Navier-Stokes formula, which results from it by setting $\alpha_1 = \alpha_2 = 0$. Like the Navier-Stokes formula, (24.2) is itself frame-indifferent, and hence it may be studied alternatively as <u>an example</u> of a fluid, for arbitrary flows. In this lecture we shall study the theory of the incompressible fluid of second grade as the Navier-Stokes theory is studied in hydrodynamics, namely, by presuming that it furnishes an exact description of the material properties of a particular fluid. The Navier-Stokes fluid and the fluid of second grade are both special Rivlin-Ericksen fluids. Like all such fluids, they are materials of the rate type, with short-range memory. It turns out that most if not all flow problems amenable to exact or approximate solution according to the Navier-Stokes theory of incompressible fluids become no more difficult when the more general fluid model is applied, but the results are entirely different.

A difference of quality is only to be expected. The only adjustable parameter in the constitutive equation of the Navier-Stokes fluid is its viscosity, μ, bearing the dimensions $ML^{-1}T^{-1}$. The fluid of second grade has also the parameters α_1 and α_2. Since the dimensions of α_i/μ are T, the fluid of second grade has a material time constant, and the magnitude and sign of such a constant can be expected to bear upon the response of the fluid to given circumstances.

Viscometric Flows.

Exercise 24.1. Show that the viscometric functions of the incompressible fluid of second grade are

$$\tau(\kappa) = \mu\kappa$$

$$\sigma_1(\kappa) = (2\alpha_1 + \alpha_2)\kappa^2 \quad , \tag{24.3}$$

$$\sigma_2(\kappa) = \alpha_2\kappa^2 \; .$$

From this result it is plain that the viscometric functions of an incompressible fluid of second grade determine, in turn, its constitutive equation. Therefore, just as for the Navier-Stokes theory, viscometric experiments suffice to discern and specify, in principle, all mechanical properties the fluid may have.

Exercise 24.2. Calculate the constitutive equation of the incompressible fluid of third grade and show that its viscometric functions do not determine it.

There are theoretical reasons to expect that

$$\mu > 0 \; , \quad \alpha_1 < 0 \; , \tag{24.4}$$

and experimental data, to the limited extent they exist, support these inequalities, which we shall accordingly adopt.

Time-dependent Shearings.

Consider the velocity field of a shearing:

$$\dot{x}_1 = 0 \; , \; \dot{x}_2 = v(x_1, t) \; , \; \dot{x}_3 = 0 \; . \tag{24.5}$$

This flow may be visualized as produced in a fluid confined between two parallel plates, x_1 = const., which are driven in some motion parallel to one another. The fluid is assumed to adhere to these plates. The steady shearings (8.8) considered in Lectures 8 and 9 are included as a special case of (24.5). Substitution of (24.5) into (24.2), followed by substitution of the result into Cauchy's first law (3.16), yields the following linear differential equation for $v(x, t)$, where we write x for x_1:

$$\mu\partial_x^2 v + \alpha_1\partial_t\partial_x^2 v + c(t) = \rho\partial_t v . \tag{24.6}$$

Exercise 24.3. Verify this result. Interpret $c(t)$ as the driving force. When $c(t) = 0$, obtain exact solutions of the form

$$v = Ve^{-ax} \cos (\omega t - bx) , \tag{24.7}$$

where V, a, b, and ω are constants. Discuss the behavior of the absorption a and the phase-shift b as functions of the frequency ω, for various fixed values of α_1.

The coefficient α_2 in (24.2) does not appear in (24.6). If $\alpha_1 = 0$, (24.6) reduces to the heat equation with source $c(t)$, just as it does for the Navier-Stokes fluid, in the theory of which the shearings (24.5) occupy a prominent position as one of the few classes of flows for which exact solutions are easily obtained. These solutions are important for illustrating the birth, growth, and decay of plane boundary layers. For a fluid of second grade such that $\alpha_1 \neq 0$, the field of speeds $v(x, t)$ differs from that gotten from the Navier-Stokes equations. More than this, the equation to be solved is of third rather than second order. As might be expected, there is a greater range of possible solutions.

Instability of Shearing Flows.

COLEMAN, DUFFIN, & MIZEL found that all shearing flows of a fluid of second grade in which the inequalities (24.4) hold are unstable in a sense we shall presently make plain. We consider only the case when $c(t) = 0$.

We observe first that the particular function

$$v = Ae^{a_n t} \sin \frac{n\pi}{d} x \; , \tag{24.8}$$

where

$$a_n = \frac{\mu^2 n^2 \pi^2}{(-\alpha_1)(n^2\pi^2 - \frac{\rho d^2}{-\alpha_1})} \; , \tag{24.9}$$

is a real solution of (24.7) which vanishes on the planes $x = 0$ and $x = d$ and which is unbounded as $t \to \infty$, provided that

$$n > \sqrt{\frac{\rho}{-\alpha_1}} \frac{d}{\pi} \; . \tag{24.10}$$

This solution represents a growing oscillation of a fluid confined by the two parallel, stationary walls $x = 0$ and $x = d$.

Suppose now that $v(x, t)$ is a bounded solution of (24.6) in the channel $0 \leqq x \leqq d$ for all time. Set

$$v^* \equiv v + \varepsilon e^{a_n t} \sin \frac{n\pi}{d} x \; . \tag{24.11}$$

Then v^* also is a solution of (24.6), and

$$|v^*(x, 0) - v(x, 0)| \leqq \varepsilon \; . \tag{24.12}$$

Moreover, $v^*(x,t)$ is unbounded as $t \to \infty$. Therefore, if for the fluid between the two stationary plates $x = 0$ and $x = d$ there exists for some particular initial speed field $v(x, 0)$) a bounded solution $v(x, t)$, for another initial speed field $v^*(x, 0)$ that differs by an arbitrarily small amount from $v(x, 0)$ there exist infinitely many unbounded solutions. That is, the kind of internal friction represented by the fluid of second grade may amplify rather than damp disturbances.

No such objection applies to the Navier-Stokes equations. If $\alpha_1 = 0$, the foregoing argument collapses.

Interpretation.

The instability theorem just proved makes it plain that the fluid of the second grade does not afford a good model for physical fluids. Nevertheless, in the last lecture we gave reasons for regarding the fluid of second grade as an improvement upon the Navier-Stokes fluid, in the sense of successive approximation. How can this be?

In fact, there is no contradiction. The Navier-Stokes theory emerges at the first stage of approximation for <u>slow</u> flows. Beyond this, it has been found for over a century to give excellent results in <u>general</u> flows of many, though not all, physical fluids. In slow flows, ultimately, the fluid of second grade provides a better approximation to a general fluid than does the Navier-Stokes fluid. This fact does not give any grounds for expecting it to be a better theory for general flows.

First, it is possible that for some particular physical fluid the Navier-Stokes theory is really the correct one. Then $\alpha_1 = \alpha_2 = 0$, and the above objection against the fluid of second grade breaks down. This kind of explanation, however, rests on a conjecture about results of experiment and thus is not to be taken seriously.

Second, the retardation theorem refers to the constitutive equation itself and gives no status to the differential equations of motion resulting from it. If $a(x) - b(x)$ is small, this does not mean that $a'(x) - b'(x)$ is small (if it did, we could conclude from the fact that the father is taller than the son that the father is also growing faster). A small term in a differential equation does not necessarily have a small effect on the solution.

Third, a process of approximation by series expansion often yields successive terms which afford better approximation in a narrow range at the expense of worse approximation in a broad one. For example, the tangent to a differentiable monotone function joins the function in being also monotone. Thus such a tangent, at any point, preserves the main property of the function being approximated, although the actual error may be great. The second approximation by Taylor's theorem is a parabola, which certainly is better than the tangent near the particular point, but far away is far worse, being in fact even of opposite sign at either $+\infty$ or $-\infty$. Likewise, the function $y = \sin x$ is fairly well approximated by any of the lines $y = k$, where $|k| \leqq 1$; although only two of these are even tangent anywhere to the curve, they all preserve two of its main properties, namely, being periodic with the same period (among others) and bounded by the same bounds. No polynomial approximation, and hence no finite number of terms in any Taylor approximation, has either of these properties. These examples show that in unbounded regions a crude first approximation may give a far better overall picture than does any higher approximation, however refined, near a single point.

The retardation theorem delivers approximation near the rest history. There is no reason to expect them to be good in general. By experience, the Navier-Stokes theory has been shown to be useful for unrestricted flows. The theory of the fluid of second grade, clearly, is not.

LECTURE 25: SECONDARY FLOWS IN PIPES

The Problem:

In Lecture 10 we have seen that steady rectilinear flow down a straight pipe is possible for a general incompressible fluid only for certain exceptional cross-sections, and for all cross-sections only for certain exceptional fluids. Both the Navier-Stokes fluid and the fluid of second grade are exceptional in this sense, since their viscometric functions satisfy the special relation (10.21). If a general fluid is forced by a pressure gradient down a pipe bounded by a cylindrical surface that is neither a pair of parallel planes nor a pair of coaxial circular cylinders, we expect that a steady flow be set up, but it cannot be rectilinear. The component of velocity normal to the generation of the cylindrical boundary is called a secondary flow. The simplest velocity field of this kind would seem to be

$$\dot{\underset{\sim}{x}} = v(\underset{\sim}{p})\underset{\sim}{k} + \underset{\sim}{u}(\underset{\sim}{p}) \ , \quad \dot{\underset{\sim}{x}} = \underset{\sim}{0} \quad \text{if } \underset{\sim}{p} \ \varepsilon \partial \mathcal{a} \ ,$$

$$\operatorname{div} \underset{\sim}{u}(\underset{\sim}{p}) = 0 \ , \tag{25.1}$$

where $\mathcal{a}$ is the plane cross-section, $\underset{\sim}{k}$ is a unit vector normal to it, $\underset{\sim}{u}(\underset{\sim}{p})$ lies in $\mathcal{a}$, and $\underset{\sim}{p}$ is the position vector in $\mathcal{a}$. Thus a solution is sought in which the secondary flow is the same at each cross-section.

At present, the calculation of secondary flows in pipes for fluids in general seems to offer an insuperable problem. We might hope to be able to show that as the driving force is made to approach 0, a family of retarded flows results, but no such theorem has been proved. For want of it, we are forced to assume the conclusion it would yield. Namely, we assume that the solution may be expressed as a power series in the specific driving force, and we assume also that the constitutive equation is sufficiently approximated by that of a Rivlin-Ericksen fluid of some sufficiently large grade.

After ERICKSEN conjectured the existence of these secondary flows in 1956, GREEN & RIVLIN immediately calculated the solution for flow in an elliptical pipe according to a particular theory. A general method was worked out later by LANGLOIS & RIVLIN. I shall outline that method, as simplified by NOLL.

<u>Preliminary Reduction</u>.

We assume that $\mathcal{a}$ is simply connected. Since $\operatorname{div} \underset{\sim}{u}(\underset{\sim}{p}) = 0$, there exists a single-valued stream function in $q(\underset{\sim}{p})$ such that

$$\underset{\sim}{u} = (\nabla q)^{\perp} \quad . \tag{25.2}$$

Here we are writing ∇ for the spatial gradient and are using the notation

$$\underset{\sim}{v}^{\perp} \equiv \underset{\sim}{k} \times \underset{\sim}{v} \tag{25.3}$$

for any vector $\underset{\sim}{v}$. On $\mathcal{a}$ the boundary condition expressing adherence of the fluid is

$$q = 0 \;, \quad \partial_n q = 0 \;, \tag{25.4}$$

where ∂_n is the normal derivative. Since the velocity is independent of the longitudinal coordinate x, so is $\underset{\sim}{T}$, and the z-component of Cauchy's first law (3.16) yields

$$\rho\phi = -\,az + \zeta(\underset{\sim}{p}) \quad , \tag{25.5}$$

where, as usual, $\phi \equiv p/\zeta + \upsilon$, υ being the potential of the body force. We shall assume that $\upsilon = \upsilon(\underset{\sim}{p})$. The function $\zeta(\underset{\sim}{p})$ remains to be determined. The constant a is the specific driving force, which we shall regard as the parameter governing the problem.

We employ the following decomposition of the stress tensor:

$$\underset{\sim}{S} \equiv \underset{\sim}{T} + p\underset{\sim}{1} = N(\underset{\sim}{k} \otimes \underset{\sim}{k}) + \underset{\sim}{t} \otimes \underset{\sim}{k} + \underset{\sim}{k} \otimes \underset{\sim}{t} + \underset{\sim}{\Pi} \;, \tag{25.6}$$

where N is a scalar, $\underset{\sim}{t}$ is a vector in $\mathcal{a}$, and $\underset{\sim}{\Pi}$ is a tensor in $\mathcal{a}$ By (25.5),

$$T\langle zz\rangle = -p + N, \qquad (25.7)$$
$$= N(\underset{\sim}{p}) + az - \zeta(\underset{\sim}{p}) + \rho\upsilon(\underset{\sim}{p}) ,$$

while the other two components of Cauchy's first law (3.16) assume the forms

$$\operatorname{div} \underset{\sim}{t} + a = \rho\underset{\sim}{u} \cdot \nabla v ,$$
$$\operatorname{div} \underset{\sim}{\Pi} - \nabla\zeta = \rho(\nabla\underset{\sim}{u})\underset{\sim}{u} . \qquad (25.8)$$

Expansions in Powers of a.

We assume that all flow quantities are expansible in powers of a with expansions having a common, non-zero interval of convergence.

$$v = \sum_{r=1}^{\infty} a^r v_r ,$$
$$\underset{\sim}{u} = \sum_{r=1}^{\infty} a^r \underset{\sim}{u}_r \quad \text{where} \quad \underset{\sim}{u}_r = (\nabla q_r)^{\perp} , \qquad (25.9)$$
$$\zeta = \sum_{r=1}^{\infty} a^r \zeta_r ,$$

the boundary conditions on $\partial\mathcal{a}$ being

$$v_r = 0 , \quad q_r = 0 , \quad \partial_n q_r = 0 . \qquad (25.10)$$

The velocity gradient $\underset{\sim}{G}$ is given by

$$\underset{\sim}{G} = \underset{\sim}{k} \otimes \nabla v + \nabla\underset{\sim}{u} ,$$
$$= \sum_{r=1}^{\infty} a^r (\underset{\sim}{k} \times \nabla v_r + \nabla\underset{\sim}{u}_r) .$$

The Rivlin-Ericksen tensors $\underset{\sim}{A}_n$ are independent of z , and

$$\mathbf{A}_k = O(a^k) \quad \text{as} \quad a \to 0 . \tag{25.12}$$

Accordingly, we may order the terms in the constitutive equation of the differential type as follows:

$$\mathbf{S} = \sum_{k=1}^{n} \mathbf{S}_k + O(a^{n+1}) , \tag{25.13}$$

where $\mathbf{S}_k = O(a^k)$.

Exercise 25.1. Prove that

$$\mathbf{S}_1 = \mu \mathbf{A}_1 ,$$

$$\mathbf{S}_2 = \alpha_1 \mathbf{A}_2 + \alpha_2 \mathbf{A}_1^2 ,$$

$$\mathbf{S}_3 = \beta_1 \mathbf{A}_3 + \beta_2 (\mathbf{A}_2 \mathbf{A}_1 + \mathbf{A}_1 \mathbf{A}_2) + \beta_3 (\text{tr } \mathbf{A}_2) \mathbf{A}_1 ,$$

$$\begin{aligned} \mathbf{S}_4 = {} & \gamma_1 \mathbf{A}_4 + \gamma_2 (\mathbf{A}_3 \mathbf{A}_1 + \mathbf{A}_1 \mathbf{A}_3) + \gamma_3 \mathbf{A}_2^2 \\ & + \gamma_4 (\mathbf{A}_2 \mathbf{A}_1^2 + \mathbf{A}_1^2 \mathbf{A}_2) \\ & + \gamma_5 (\text{tr } \mathbf{A}_2) \mathbf{A}_2 + \gamma_6 (\text{tr } \mathbf{A}_2) \mathbf{A}_1^2 \\ & + [\gamma_7 \text{tr } \mathbf{A}_3 + \gamma_8 \text{tr}(\mathbf{A}_2 \mathbf{A}_1)] \mathbf{A}_1 , \end{aligned} \tag{25.14}$$

where $\alpha_1, \alpha_2, \beta_1, \beta_2, \beta_3, \gamma_1, \ldots \gamma_8$ are constants. In general $\sum_{k=1}^{n} \mathbf{S}_k$ gives the constitutive equation of an incompressible Rivlin-Ericksen fluid of grade n.

The expansion (25.13), the first four terms of which are written out as (25.14), is set alongside (25.9), and all are substituted into (25.8). Equating to 0 the coefficients of the successive powers of a then yields equations for determining the functions $v_r(\mathbf{p})$ and $\mathbf{u}_r(\mathbf{p})$ successively.

We shall outline the calculation of the first four steps in the process.

Step 1. The Navier-Stokes Solution.

Exercise 25.2. Prove that

$$\mu\Delta v_1 = -1 ,$$

$$\mu \operatorname{div}[\nabla \underset{\sim}{u}_1 + (\nabla \underset{\sim}{u}_1)^T] = \nabla\zeta_1 , \tag{25.15}$$

so that

$$\mu\Delta\Delta q_1 = 0 . \tag{25.16}$$

The Navier-Stokes solution of the problem is given by the unique function v_1 that satisfies $(25.15)_1$ and the boundary condition $v_1 = 0$ if $\underset{\sim}{p}\varepsilon\partial\mathcal{a}$. Since (25.16) is a biharmonic equation, it has a unique solution q_1 such that q_1 and $\partial_n q_1$ vanish on $\partial\mathcal{a}$. That solution is $q_1 \equiv 0$. Hence the first-order solution is the Navier-Stokes solution:

$$\dot{x} = av_1\underset{\sim}{k} + O(a^2) . \tag{25.17}$$

Step 2. Solution for the Fluid of Second Grade.

Exercise 25.3. Setting $x_1^2 \equiv (\nabla v_1)^2$, show that

$$\underset{\sim}{A}_1^2 = a^2[\nabla v_1 \otimes \nabla v_1] + \kappa_1^2(\underset{\sim}{k} \otimes \underset{\sim}{k}) + O(a^3) ,$$

$$\underset{\sim}{A}_2 = 2a^2(\nabla v_1 \otimes \nabla v_1) + O(a^3) , \tag{25.18}$$

and hence

$$\mu\Delta v_2 = 0 , \quad \mu\Delta\Delta q_2 = 0 , \tag{25.19}$$

so that $v_2 = 0$, $q_2 = 0$, $\underset{\sim}{u}_2 = \underset{\sim}{0}$, and

$$\dot{\underset{\sim}{x}} = av_1\underset{\sim}{k} + O(a^3) . \tag{25.20}$$

That is, the Navier-Stokes velocity field is the exact solution also for the fluid of second grade. Of course, normal tractions not included in the Navier-Stokes theory are required to produce the same flow in the fluid of second grade.

Step 3. Solution for the Fluid of Third Grade.

Exercise 25.4. Show that $\underset{\sim}{A}_3 = O(a^5)$ and hence

$$\underset{\sim}{t} = a\mu\nabla v_1 + a^3[\mu\nabla v_3 + 2(\beta_2 + \beta_3)\kappa_1^2 \nabla v_1] + O(a^4) ,$$

$$\underset{\sim}{\Pi} = a^2(2\alpha_1 + \alpha_2)(\nabla v_1 \otimes \nabla v_1) + a^3\mu[\nabla \underset{\sim}{u}_3 + (\nabla \underset{\sim}{u}_3)^T] + O(a^4) . \quad (25.21)$$

Substitution in (25.8) yields

$$\mu\Delta v_3 = -2(\beta_2 + \beta_3)\operatorname{div}[(\nabla v_1)^2\nabla v_1] ,$$

$$\mu\Delta\Delta q_3 = 0 . \quad (25.22)$$

The results of this exercise show that $q_3 = 0$, so that in the fluid of third grade there is no secondary flow. The function v_3 is uniquely determined by the Poisson equation $(25.21)_1$ in terms of the Navier-Stokes flow field v_1, the viscosity μ, and the third-order viscosities β_2 and β_3. Thus

$$\dot{x} = (av_1 + a^3v_3)\underset{\sim}{k} + O(a^4) \quad (25.23)$$

where v_1 and v_3 are functions of $\underset{\sim}{p}$ depending also on μ, β_2, and β_3, for any given cross-section $\mathcal{a}$.

Step 4. Secondary Flow for the Fluid of Fourth Grade.

Exercise 25.5. Prove that

$$\underset{\sim}{t} = a(\ldots) + a^3(\ldots) + a^4\mu\nabla v_4 + O(a^5) ,$$

$$\underset{\sim}{\Pi} = a^2(\ldots) + a^4\{\mu[\nabla\underset{\sim}{u}_4 + (\nabla\underset{\sim}{u}_4)^T] + (2\alpha_1 + \alpha_2)[\nabla v_3 \otimes \nabla v_1 + \nabla v_1 \otimes \nabla v_3] + \gamma\kappa_1^2 \nabla v_1 \otimes \nabla v_1\} + O(a^5) , \quad (25.24)$$

where

$$\gamma \equiv 4(\gamma_3 + \gamma_4 + \gamma_5 + \frac{1}{2}\gamma_6) \ . \tag{25.25}$$

Substitution in (25.8) yields

$$\begin{gathered} \mu \Delta v_4 = 0 \ , \\ \mu \operatorname{div} [\nabla \underset{\sim}{u}_4 + (\nabla \underset{\sim}{u}_4)^T] \\ + (2\alpha_1 + \alpha_2) \operatorname{div}[\ v_3 \otimes \nabla v_1 + \nabla v_1 \otimes \nabla v_3] \\ + \gamma \operatorname{div}[\kappa_1^2 \ \nabla v_1 \otimes \nabla v_1] - \nabla \zeta_4 = \underset{\sim}{0}. \end{gathered} \tag{25.26}$$

Therefore $v_4 = 0$. Simplify $(25.26)_2$ so as to obtain

$$\mu \Delta\Delta q_4 = \delta (\nabla v_1)^{\perp} \cdot \nabla \operatorname{div}[(\nabla v_1)^2 \nabla v_1] \ , \tag{25.27}$$

where

$$\delta \equiv \gamma - \frac{2}{\mu}(2\alpha_1 + \alpha_2)(\beta_2 + \beta_3) \ . \tag{25.28}$$

Since (25.27) is an inhomogeneous biharmonic equation for q_4, it has a unique solution, determined by the Navier-Stokes speed v_1 and by the constant δ, which is a particular combination of all but 5 of the 14 viscosities of orders 1, 2, 3, and 4. Except for fluids in which $\delta = 0$, or for cross-sections such that $(\nabla v_1)^{\perp} \cdot \nabla \operatorname{div}[(\nabla v_1)^2 \nabla v_1] = 0$, the solution q_4 of (25.27) will not vanish identically. Therefore, in general, a secondary flow appears at the fourth step in the expansion process outlined here. The result, then, is

$$\dot{\underset{\sim}{x}} = (a v_1 + a^3 v_3)\underset{\sim}{k} + a^4 (\nabla q_4)^{\perp} + O(a^5) \ , \tag{25.29}$$

where none of the terms written down vanishes except for special fluids or special cross-sections.

Summary

Despite the length of the calculation, the results obtained are simple. v_1 is the Navier-Stokes field of speed, uniquely determined by the Poisson equation $(25.15)_1$ and the boundary condition $v_1 = 0$ on $\partial\mathcal{a}$. Once we have v_1, we easily determine v_3 by the Poisson equation $(25.22)_1$, with the boundary condition $v_3 = 0$ on $\partial\mathcal{a}$. However, if our interest lies in the secondary flow alone, we may pass directly to $\underset{\sim}{u}_4$, the stream function of which is obtained by solving the inhomogeneous biharmonic equation (25.27) with the boundary condition $q_4 = 0$, $\partial_n q_4 = 0$ on $\partial\mathcal{a}$.

Let $\bar{v}_1$, $\bar{v}_3$, and $\bar{q}_4$ denote the solutions for the fluid such that $\mu = 1$, $\beta_2 + \beta_3 = 1$, $\delta = 1$, in some system of units. Then the solutions for the fluid with 14 arbitrary viscosities, μ, α_1, α_2, β_1, β_2, β_3, $\gamma_1, \ldots, \gamma_8$ are given by

$$v_1 = \frac{1}{\mu}\bar{v}_1 , \quad v_3 = \frac{\beta_2 + \beta_3}{\mu^4}\bar{v}_3 , \quad q_4 = \frac{\delta}{\mu^5}\bar{q}_4 . \qquad (25.30)$$

The streamline pattern of the secondary flow is given by the curves q_4 = constant. Since these are the same as the curves $\bar{q}_4$ = constant, the secondary flow pattern is the same for all fluids in which $\mu\delta \neq 0$. As is clear from (25.29), the flow is generally of the helical kind, the particles travelling down the tube in some sort of spirals. The pitch of these spirals, unlike the secondary pattern alone, varies from one fluid to another and depends also upon a. Of course the distribution of speeds along the streamlines and the tractions necessary to produce the flow depend also on the particular flow.

In Lecture 10 we have seen that a rectilinear flow in a pipe is a viscometric flow but generally is not possible unless certain non-conservative body force is supplied. The foregoing analysis determines the flow, to within terms $O(a^5)$, that does occur subject to the driving force a alone. It is not a viscometric flow. Nevertheless, as the results of the following exercise show, this flow is completely determined by the viscometric functions of the fluid.

<u>Exercise 25.6</u>. Show that for the fluid of fourth grade

$$\tilde{\mu}(\kappa) = \mu + 2(\beta_2 + \beta_3)\kappa^2 \quad ,$$

$$\sigma_1(\kappa) = (2\alpha_1 + \alpha_2)\kappa^2 + [4(\gamma_3 + \gamma_4 + \gamma_5) + 2\gamma_6]\kappa^4 \quad , \qquad (25.31)$$

$$\sigma_2(\kappa) = \alpha_2\kappa^2 + 2\gamma_6\kappa^4 \quad .$$

Hence v_3, the perturbation of the longitudinal speed, is determined for any given cross-section by the quantity

$$\frac{\tilde{\mu}''(0)}{2\mu}$$

and the quantity δ determining the secondary flow is given by

$$\delta = \frac{\sigma_1^{\,iv}(0)}{24} - \frac{\sigma_1''(0)\tilde{\mu}''(0)}{4\mu} \quad . \qquad (25.32)$$

The results thus have a double importance. First, the flow pattern is different from that predicted by the classical Navier-Stokes theory. Second, information gotten in viscometric flows is sufficient to predict the non-viscometric flow that does occur. Of course, for the Navier-Stokes fluid and the fluid of second grade, <u>all</u> flows are determined by viscometric information, and originally the aim of viscometry was to ascertain the nature of the fluid and hence to determine, in principle, all flows it might experience. For general fluids, the secondary flow in pipes affords the only case so far known in which the results of experiments on viscometric flows serve to predict a phenomenon falling outside the viscometric area. We notice that the function $\sigma_2(\kappa)$, which has no effect on the possibility or impossibility of rectilinear flow, also has no effect on the flow that does occur according to the theory of the fluid of fourth grade.

Since the theory of the partial differential equations found is standard, we may justly regard the problem of calculating secondary flows according to the theory of the fluid of fourth grade as solved by the foregoing analysis. We illustrate the results by the example of the elliptical pipe.

Example: The Elliptical Cross-Section.

For the ellipse

$$\frac{x^2}{c^2} + \frac{y^2}{b^2} = 1 \ , \qquad c > b \ , \tag{25.33}$$

the classical solution is

$$v_1 = - \frac{c^2 b^2}{2\mu(c^2 + b^2)} \left(\frac{x^2}{c^2} + \frac{y^2}{b^2} - 1\right) . \tag{25.34}$$

Exercise 25.7. Show that (25.27) becomes

$$\mu^5 \Delta\Delta q_4 = \delta \frac{6c^2b^2(c^2 - b^2)}{(c^2 + b^2)} xy \ . \tag{25.35}$$

Show that the stream function of the secondary flow is

$$q_4 = \frac{\delta}{\mu^5} A\left(\frac{x^2}{c^2} + \frac{y^2}{b^2} - 1\right)^2 xy \ , \tag{25.36}$$

where

$$A = \frac{c^6 b^6 (c^2 - b^2)}{4(c^2 + b^2)^3 (5c^4 + 6c^2b^2 + 5b^4)} . \tag{25.37}$$

Determine and discuss the streamline pattern.

The solution given by (25.36) is that first obtained by GREEN & RIVLIN for a special theory.

Remarks.

It is often objected that perturbation methods are not really enlightening because, since they never stray far from what was known before, they are fit to describe only situations that are essentially classical.

The foregoing problem affords one of the rare cases in which a perturbation method has delivered something really new. As $a \to 0$, the resulting velocity field approaches the classical one, it is true. However, the presence of any steady component, however small, of velocity normal to the main flow yields helical stream lines. Therefore, the particles are <u>not</u> made to remain arbitrarily close to their classical positions by making a arbitrarily small. True, the smaller is a, the greater length of pipe will have to be inspected in order to ascertain the difference from the classical prediction, but if we peer down the infinite pipe, we shall see the projections of the streamlines as closed curves whose forms are <u>independent</u> of a.

Most of the more striking phenomena of "non-linear viscosity" were noticed first in nature and then later found to be easily covered by any fairly general theory. ERICKSEN's secondary flow in non-circular pipes is the most important of the phenomena predicted by theory before being observed in experiment.

PART V. THERMODYNAMICS

LECTURE 26: THERMODYNAMICS OF HOMOGENEOUS PROCESS.

Temperature and Heat.

Common experience makes it plain that mechanical action does not always give rise to mechanical effects alone. Doing work upon a body may make it hotter, and heating a body may cause it to do work. This fact suggests that alongside the mechanical working or power P should be set a heat working Q, such that the total working is $P + Q$. If kinetic energy is the energy of motion, we recognize that not all work leads to motion and allow for storage of the remainder as internal energy E. Thus the balance of energy is asserted in the equation

$$\dot{K} + \dot{E} = P + Q \ , \tag{26.1}$$

where the dots denote time rates.

Much as the concept of place abstracts the idea of where a body is, the concept of temperature abstracts the idea of how hot it is. The temperature θ is a real number indicated on a thermometer, just as distance is a real number indicated on a ruler. Experience suggests that no matter what thermometer is used, there is a temperature below which no body can be cooled. That is, the temperature is bounded below. If this greatest lower bounded is assigned the temperature 0, the temperature is said to be absolute:

$$\theta > 0 \ , \tag{26.2}$$

no matter what unit of temperature be selected. We shall always adopt (26.2) for convenience.

Heat is not to be confused with internal energy E. Internal energy is stored when work is done without producing motion, and it may be reconverted into motion or working, as (26.1) asserts. In common experience work done generally raises the temperature of a body, but a decrease of temperature does not necessarily release the energy so gained. A somewhat broader idea is expressed by assuming the existence of an assignable bound B for the heat working:

$$Q \leqq B \; . \tag{26.3}$$

Equivalently, by (26.1), $\dot{E} + (\dot{K} - P)$ is bounded by B.

In terms of the bound B, the heat H is defined by

$$H \equiv \int \frac{B}{\theta}\, dt \; . \tag{26.4}$$

Thus the heat is defined only to within an arbitrary constant. The heat is of importance for the theory only through $\dot{H}$, the heating. In terms of it, (26.3) assumes the form

$$Q \leqq \theta \dot{H} \; . \tag{26.5}$$

A body stores heat, then, at least as fast as the heat working divided by the temperature. The hotter the body is, the greater must be the heat working in order to reach the maximum, which is determined by the heating.

Remarks on the Experiential Basis of Thermodynamics.

The student of continuum mechanics may be expected to have had a more or less sound introduction to the simple mechanics of forces and "bodies" of unspecified nature. He is unlikely to have had any such training with heat and temperature. While these concepts are just as close to common physical experience as are place, time, mass, and force, their mathematical theory has not been put into a simple, explicit, and concise form like mechanics. Most books on thermodynamics devote a lot of space to motivation and to correspondence between theory and experiment, which an elementary book on mechanics generally omits in favor of concrete mathematics. While in mechanics the nature of measurement of time, place, mass, and force is left to works on the philosophy or foundations of physics or relativity, the corresponding aspects of thermodynamics are treated at length in every book, and badly. The present lecture is a concession to this state of affairs.

The intended application of elementary thermodynamics is to blocks of material suffering no local deformation or variation of temperature. We describe the theory, therefore, as one of homogeneous processes in that all quantities occurring are functions of time only. Place plays no part. It is customary in the older studies to neglect motion and hence to set $K = 0$, but for later use we shall retain K, noting, however, that it occurs in the theory of homogeneous processes only in the combination $P - \dot{K}$. Since this quantity is the excess of the mechanical working over the kinetic energy rate, it may be called the work storage.

I think it unnecessary to supply motivation beyond that already given. For connection with other treatments, however, I remark that what I call heat H is usually called "entropy"; that (26.1) and (26.5), respectively, are among the various statements that physical writers call the "first and second laws of thermodynamics"; that, specifically, (26.5) is the Clausius-Planck inequality.

I shall now outline a mathematical theory of the thermodynamics of homogeneous processes, in the same way as I should present elementary mechanics to a beginning class, as a clean, self-contained mathematical theory, in which general theories may be proved and specific problems may be solved.

The Thermodynamic State

Much as in mechanics the concept of place occupied at a given time is primitive, in thermodynamics we consider a state at a given time. The state is one set of $k + 1$ functions of time

$$\theta(t) \ , \ T_1(t), \ldots, \ T_k(t)$$

or briefly,

$$\theta(t) \ , \ T(t)$$

The temperature $\theta(t)$ is a positive function: $\theta(t) > 0$. The k parameters $T_j(t)$ also are quantities given a priori. In the interpretation, T_1 might be the volume of a body, and $T_2, \ldots, T_k$ might be the masses of $k-1$ constituents, but for the theory, no specific interpretation is necessary. We agree to consider only unconstrained states in the sense that the functions

$\theta(t)$, $T(t)$ may be arbitrary. In particular, at any one time θ, T, $\dot{\theta}$, and $\dot{T}$ may assume any real values whatever, provided only that $\theta > 0$.

Thermodynamic Processes.

A thermodynamic process is a set of $k + 5$ functions

$$\theta(t)\ ,\quad T(t)\ ,\quad P(t) - \dot{K}(t)\ ,\ E(t)\ ,\ Q(t)\ ,\ H(t)\ ,$$

that satisfy the first two axioms of thermodynamics:

$$\dot{K} + \dot{E} = P + Q \tag{26.1}$$

$$\theta > 0 \tag{26.2}$$

If all but the last two of these functions are given, $Q(t)$ is uniquely determined by (26.1). Hence it is equivalent to say that a thermodynamic process is a set of $k + 3$ functions

$$\theta(t)\ ,\ T(t)\ ,\ P(t) - \dot{K}(t)\ ,\ E(t)\ ,\ H(t)\ .$$

such that $\theta > 0$.

If we use (26.1) to eliminate Q from (26.5), we obtain the reduced dissipation inequality:

$$\dot{E} \leqq P - \dot{K} + \theta\dot{H}\ . \tag{26.6}$$

The free energy ψ is defined by

$$\psi \equiv E - \theta H\ . \tag{26.7}$$

In terms of it, the reduced dissipation inequality becomes

$$\dot{\psi} - (P - \dot{K}) + H\dot{\theta} \leqq 0\ . \tag{26.8}$$

A process in which

$$\left.\begin{array}{l}\dot{\theta} = 0 \\ Q = 0 \\ \dot{H} = 0 \\ \dot{E} = 0\end{array}\right\} \quad \text{is called} \quad \left\{\begin{array}{l}\text{isothermal} \\ \text{adiabatic} \\ \text{isentropic} \\ \text{isoenergetic}\end{array}\right.$$

A process in which equality holds in (26.8) at all time is called reversible. All other processes are called irreversible. In a reversible process, $Q = \theta\dot{H}$; therefore, in reversible process "adiabatic" and "isentropic" have the same meaning. In an irreversible adiabatic process, the heat increases: $\dot{H} > 0$. In an irreversible isentropic process, the heat working is negative: $Q < 0$. So much for terms.

Thermodynamic Constitutive Equations.

The state history is the set of $k + 1$ functions $\theta^t(s)$, $T_k^t(s)$. The principle of thermodynamic determinism asserts that the state history determines the work storage, the free energy, and the heat. That is,

$$\left.\begin{array}{c}P-\dot{K} \\ \Psi \\ H\end{array}\right\} \quad \text{are functionals of} \quad \left\{\begin{array}{c}\theta^t \\ \\ T^t\end{array}\right. \tag{26.9}$$

Such a set of three relations is called a thermodynamic constitutive equation. Not all functionals are admissible, however. In order to define a thermodynamic material, we require that the constitutive equation be compatible with the reduced dissipation inequality (26.8) for all thermodynamic processes.

Example 1. The classical caloric equations of state.

The classical works in thermodynanics deal only, or almost only, with one extremely special class of constitutive equations. The work storage is thought of as being purely mechanical and also linear in the "velocities" $\dot{T}_a$.

$$P - \dot{K} = - \sum_{a-1}^{k} \tilde{\omega}_a(\theta, T)\dot{T}_a \quad , \tag{26.10}$$

where the coefficients $\tilde{\omega}_a$ are called thermodynamic pressures, and the free energy and heat are given by caloric equations of state:

$$\Psi = \Psi(\theta, T) \;, \qquad H = H(\theta, T) \;. \tag{26.11}$$

Substitution of (26.10) and (26.11) into the reduced dissipation inequality (26.8) yields

$$(H + \partial_\theta \Psi)\dot{\theta} + \sum_{a=1}^{k} (\tilde{\omega}_a + \partial_{T_a} \Psi)\dot{T}_a \leqq 0, \tag{26.12}$$

The coefficients of $\dot{\theta}$ and of $\dot{T}_a$ are functions of θ and T. Since θ, T, $\dot{\theta}$, $\dot{T}$, may be given arbitrary values, (26.12) holds if and only if each summand vanishes:

$$H = -\partial_\theta \Psi \,, \qquad \tilde{\omega}_a = -\partial_{T_a} \Psi \;. \tag{26.13}$$

We have proved then, that the $k + 2$ functions $\tilde{\omega}_a$, T, and H are not independent in a thermodynamic constitutive equation: The heat and the thermodynamic pressures are determined from the free-energy function $\Psi(\theta, T)$ as a potential. Substitution back into (26.12) shows that equality holds always there. That is, there are no irreversible processes, and a process is adiabatic if and only if it is isentropic.

The contents of classical books on thermodynamics consists mainly in ringing the changes on (26.13).

Exercise 26.1. Assume that

$$P - \dot{K} = -\sum_{a=1}^{k} \tilde{\omega}_a \dot{T}_a \tag{26.14}$$

without necessarily imposing the assumption that $\tilde{\omega}_a = \tilde{\omega}_a(\theta, T)$. I.e., the $\tilde{\omega}_a$ are certain functions of t. Define the enthalpy X and the free enthalpy Z as follows:

$$X = E + \sum_{a=1}^{k} \tilde{\omega}_a T_a \; ,$$

$$Z = X - \theta H \; , \qquad (26.15)$$

so that

$$E + Z = \psi + X \; . \qquad (26.16)$$

Prove that

$$\dot{X} \leqq \theta \dot{H} + \sum_{a=1}^{k} \dot{\tilde{\omega}}_a T_a \; ,$$

$$\dot{Z} \leqq -H\dot{\theta} + \sum_{a=1}^{k} \dot{\tilde{\omega}}_a T_a \; , \qquad (26.17)$$

and interpret these inequalities.

Exercise 26.2. Under the conditions of the theorem expressed by (26.13), assume that each of the functions $\psi(\theta,T)$, $H(\theta,T)$, $\tilde{\omega}_a(\theta,T)$ is invertible for any one of its arguments. Prove that then $E = E(H,T)$, $X = X(H,\tilde{\omega})$, and $Z = Z(\theta,\tilde{\omega})$, and each of these three functions is also a thermodynamic potential in the sense that

$$\begin{aligned} \theta &= \partial_H E \; , & \tilde{\omega}_a &= -\partial_{T_a} E \; , \\ \theta &= \partial_H X \; , & T_a &= \partial_{\tilde{\omega}_a} X \; , \\ H &= -\partial_\theta Z \; , & T_a &= \partial_{\tilde{\omega}_a} Z \; . \end{aligned} \qquad (26.18)$$

The existence of the classical potentials, as we have just seen, is a consequence, not of any general thermodynamic ideas, but of the very special constitutive equations (26.10) and (26.11) taken as the starting point. If we generalize these assumptions just a little, we no longer get the classical formulae. Namely, if we include the effect of "linear friction", the pressure $\tilde{\omega}_a$ in (26.10) is no longer a function of θ and T alone but is also a linear function of $\dot{T}$ and $\dot{\theta}$:

$$\tilde{\omega}_a = \tilde{\omega}_a^0(\theta,T) + \sum_{b=1}^{k} \tilde{\omega}_{ab}(\theta,T)\dot{T}_b \; , \qquad (26.19)$$

while the caloric equations of state (26.11) are retained unchanged. Here $\tilde{\omega}^0_a(\theta,T)$ is the "equilibrium pressure", namely, the value of $\tilde{\omega}_a$ when $\dot{T} = 0$. Substitution into the reduced dissipation inequality (26.8) yields now

$$(H + \partial_\theta\psi)\dot{\theta} + \sum_{a=1}^{k} (\tilde{\omega}^0_a + \partial_{T_a}\psi)\dot{T}_a + \sum_{a,b=1}^{k} \tilde{\omega}_{ab}\dot{T}_a\dot{T}_b \leqq 0 . \tag{26.20}$$

We shall analyse this inequality somewhat more carefully. Since the coefficients of $\dot{\theta}$, $\dot{T}_a$, and $\dot{T}_a\dot{T}_b$, are functions of θ and T only, and since we may construct a thermodynamic process in which the $2k + 2$ quantities θ, T_a, $\dot{\theta}$, $\dot{T}_b$ have any values we choose to assign at the time t, provided only that $\theta > 0$, the inequality (26.20) is of the type

$$\sum_i A_i x_i + \sum_{i,j} A_{ij} x_i x_j \leqq 0 \tag{26.21}$$

for all x_k, where the A_i and A_{ij} are constants. First, let the x_k be particular numbers; then (26.1) must hold if they are replaced by αx_k, where α is any constant. Therefore, for all α

$$\alpha\Sigma A_i x_i + \alpha^2 \Sigma A_{ij} x_i x_j \leqq 0 . \tag{26.22}$$

This quadratic inequality is satisfied for all α if and only if the coefficient of α is 0 while the coefficient of α^2 is non-positive. That is

$$\Sigma A_i x_i = 0 \quad , \quad \Sigma A_{ij} x_i x_j \leqq 0 . \tag{26.23}$$

These inequalities are to be satisfied for all x_i. Hence

$$A_i = 0 . \tag{26.24}$$

Conversely, if (26.24) and $(26.23)_2$ hold, (26.21) is satisfied. Applying these results to (26.20), we see that that inequality is equivalent to

$$H = -\partial_\theta \psi ,$$

$$\tilde{\omega}_a^0 = -\partial_{T_a} \psi , \qquad (26.25)$$

$$\sum_{a,b=1}^{k} \tilde{\omega}_{ab}(\theta,T)\dot{T}_a\dot{T}_b \leqq 0 .$$

Collecting our results, we see that $\psi(\theta,T)$ is not a potential for H and the equilibrium pressure $\tilde{\omega}_a^0$ but stands in no relation to the frictional pressures $\tilde{\omega}_{ab}$. The free energy function, then, no longer determines all constitutive equations of the material, since the frictional pressure $\tilde{\omega}_{ab}$ is independent of it. The frictional pressures, however, are not arbitray. For each value of the thermodynamic state, the quadratic form of $||\omega_{ab}||$ must be non-positive definite. Conversely, if (26.25) holds, the assumed constitutive equations satisfy the reduced dissipation inequality. Therefore, equality need not hold in $(26.25)_3$; indeed generally it will not hold. Thus the constitutive equations (26.19), only slightly more general than the classical (26.10), allow irreversible processes and in fact generally lead to them. With (26.19), adiabatic process are no longer generally isentropic, and conversely.

Remark.

The approach to thermodynamics presented here is due to COLEMAN & NOLL. They employed it for a certain theory of thermoelastic deformation with linear viscosity. I have presented their ideas without the complication of changes of shape and variations of temperature. In the next lecture, I shall put these ideas back in, with reference to a far more general class of constitutive equations.

LECTURE 27: THERMODYNAMICS OF SIMPLE MATERIALS.

We shall now combine the simple ideas of thermodynamics, presented in the last lecture, with those of continuum mechanics as developed in the beginning of this course. The finite set of parameters $T_1, T_2, \ldots, T_k$ is replaced by the <u>motion</u> $\underset{\sim}{\chi}(X, t)$ of a body $\mathcal{B}$; the temperature $\theta(t)$ is replaced by the <u>temperature field</u> $\theta(\underset{\sim}{x},t)$, defined over the configuration $\mathcal{B}_{\underset{\sim}{\chi}}$ of $\mathcal{B}$ at time t. We assume that $\theta(\underset{\sim}{x},t) > 0$.

<u>Energy and Heat Working in Mechanics</u>.

The laws of continuum mechanics, presented in Lecture 3, are carried over unchanged to the more general situation in which account is taken of temperature and of its variation.

The first axiom of thermodynamics, the <u>balance of energy</u>, is also unchanged:

$$\dot{K} + \dot{E} = P + Q , \qquad (26.1)$$

except that now K and P have specific forms in terms of the motion and the forces that produce it. Namely, K is the kinetic energy of the part $\mathcal{P}$ of the body $\mathcal{B}$ in the configuration $\underset{\sim}{\chi}$:

$$K \equiv \frac{1}{2} \int_{\mathcal{P}_{\underset{\sim}{\chi}}} \dot{x}^2 dm , \qquad (27.1)$$

and P, the mechanical working, is the rate at which the body force $\underset{\sim}{b}$ and the traction $\underset{\sim}{t}$ do work. If the tractions are simple, then,

$$P = \int_{\partial \mathcal{P}_{\chi}} \dot{\mathbf{x}} \cdot \mathbf{T}\mathbf{n} ds + \int_{\mathcal{P}_{\chi}} \dot{\mathbf{x}} \cdot \mathbf{b} dm , \tag{27.2}$$

where $\mathbf{T}$ is the stress tensor. The forces here are exactly the same as those contributing to the change of momentum, as expressed by (3.12).

Exercise 27.1. By use of (3.14) and Cauchy's laws (3.16) and (3.18), show that the work storage is given by

$$P - \dot{K} = \int_{\mathcal{P}_{\chi}} w dv , \tag{27.3}$$

where w is the stress working:

$$w \equiv \mathrm{tr}(\mathbf{T}\mathbf{D}) . \tag{4.17}$$

The internal energy E is taken to be an additive set function, an absolutely continuous function of mass:

$$E = \int_{\mathcal{P}_{\chi}} \varepsilon \, dm \tag{27.4}$$

where ε , the specific internal energy, is a frame-indifferent scalar:

$$\varepsilon = \varepsilon(\mathbf{x}, t) . \tag{27.5}$$

Finally, the heat working Q is regarded as arising from surface and volume densities:

$$Q = \int_{\partial \mathcal{P}_{\chi}} q ds + \int_{\mathcal{P}_{\chi}} s dm \tag{27.6}$$

(*cf.* the parallel assumption (3.2) for the resultant force). Both q and s are assumed to be frame-indifferent. The supply s is called the supply of heat working, and q is called the efflux of heat working. In this course we shall consider only external sources of heat working and simple effluxes of heat working:

$$s = s(\underset{\sim}{x}, t) ,$$

$$q = q(\underset{\sim}{x}, t, \underset{\sim}{n}) , \qquad (27.7)$$

where $\underset{\sim}{n}$ is the outer normal to $\partial \mathcal{P}_{\underset{\sim}{\chi}}$. (Cf. the parallel assumptions (3.9) and (3.10) for forces.)

Exercise 27.2. By use of $(27.6)_2$, (27.5), (27.3), and (26.1), prove FOURIER's heat flux principle: There exists a vector $\underset{\sim}{h}(\underset{\sim}{x},t)$, called the heat efflux vector, such that

$$q(\underset{\sim}{x}, t, \underset{\sim}{n}) = \underset{\sim}{h}(\underset{\sim}{x}, t)\cdot\underset{\sim}{n} \qquad (27.8)$$

(cf. CAUCHY's fundamental lemma (3.11)).

Exercise 27.3. Show from (27.8), (27.3), (4.17), and (3.13) that (26.1) is equivalent under sufficient conditions of smoothness to the differential equation

$$\rho\dot{\varepsilon} = w + \operatorname{div} \underset{\sim}{h} + \rho s \qquad (27.9)$$

(the FOURIER-KIRCHHOFF-C. NEUMANN equation of balance of energy).

Heat and Dissipation.

The heat in a body, naturally, is the sum of the heats in its parts. We regard heat as associated with mass in the sense that a massless region can hold no heat. Thus we assume that the heat H is an additive set function, absolutely continuous with respect to mass:

$$H = \int_{\mathcal{P}_{\underset{\sim}{\chi}}} \eta dm , \qquad (27.10)$$

where $\eta(\underset{\sim}{x}, t)$ for historical reasons is called the specific entropy. (The "specific heat," traditionally, is $\theta\dot{\eta}/\dot{\theta}$.)

So far, everything has been straightforward. Using the methods and concepts of continuum mechanics, we have simply carried over to fields the ideas of the thermodynamics of homogeneous processes. With the CLAUSIUS-PLANCK inequality, viz

$$Q \leqq \theta\dot{H} , \qquad (26.5)$$

several possibilities suggest themselves. One is that a corresponding relation be assumed for the densities:

$$\rho\theta\dot{\eta} \geqq \operatorname{div} \underset{\sim}{h} + \rho s \ . \tag{27.11}$$

This relation, also, we shall call the <u>Clausius-Planck inequality</u>. A second possibility is to interpret (26.5) as stating that all kinds of heat working divided by temperature give rise to heat:

$$\dot{H} \geqq \int_{\partial P_{\underset{\sim}{\chi}}} \frac{q}{\theta}\, ds + \int_{P_{\underset{\sim}{\chi}}} \frac{s}{\theta}\, dm \ . \tag{27.12}$$

The local equivalent, of course, is

$$\rho\theta\dot{\eta} \geqq \theta \operatorname{div}\left(\frac{\underset{\sim}{h}}{\theta}\right) + \rho s \ . \tag{27.13}$$

We shall refer to (27.12) and (27.13) as the <u>Clausius-Duhem inequality</u> or the <u>dissipation principle</u>. We shall lay down this principle as the generalization of (26.5) to deformable continua, subject to local variations of temperature.

If θ is uniform, both (27.13) and (27.11) reduce to (26.5). More generally, the classical heat-conduction inequality asserts that heat never flows against a temperature gradient:

$$\frac{\underset{\sim}{h}\cdot\underset{\sim}{g}}{\theta} \geqq 0 \ , \tag{27.14}$$

where

$$\underset{\sim}{g} \equiv \operatorname{grad} \theta . \tag{27.15}$$

We shall not adopt (27.14), for in fact heat may flow against a temperature gradient, just as water may flow uphill. Sufficient local supply of energy may certainly turn its flow toward any desired direction, and since such supplies are allowed here (27.14) cannot generally hold.

<u>The Reduced Dissipation Inequality</u>.

The <u>specific free energy</u> ψ is defined by analogy to (26.7):

$$\psi \equiv \varepsilon - \eta\theta \ . \tag{27.16}$$

The <u>internal dissipation</u> δ is defined by

$$\delta \equiv w - \rho(\dot{\psi} + \eta\dot{\theta}) \ , \tag{27.17}$$

where w is the stress working (4.17). Thus

$$\begin{aligned} \delta &= w - \rho(\dot{\varepsilon} - \theta\dot{\eta}) \\ &= \rho\theta\dot{\eta} - \operatorname{div} \underset{\sim}{h} - \rho s \ , \end{aligned} \tag{27.18}$$

by (27.9). This formula shows that the internal dissipation is the local excess of the rate of increase of heat per unit temperature over the heat working.

<u>Exercise 27.4</u>. Show that in a Navier-Stokes fluid (23.17), if the pressure $p(\rho)$ and the specific entropy η are obtained from the specific free energy by formulae analogous to (26.25), <u>viz</u>

$$p(\rho) = \tilde{\omega}(\upsilon) = -\partial_{\upsilon}\psi \ , \quad \eta = -\partial_{\theta}\psi \ , \tag{27.19}$$

where $\psi = \psi(\upsilon,\theta)$ and υ is the specific volume $1/\rho$, then

$$\delta = \lambda(\operatorname{tr} \underset{\sim}{D})^2 + 2\mu \operatorname{tr} \underset{\sim}{D}^2 . \tag{27.20}$$

Show that the <u>Duhem-Stokes inequalities</u>:

$$\mu \geqq 0 \ , \ 3\lambda + 2\mu \geqq 0 \ , \tag{27.21}$$

are necessary and sufficient that $\delta \geqq 0$ for all $\underset{\sim}{D}$.

Comparison with (27.11) shows that the <u>Clausius-Planck inequality holds if and only if the internal dissipation is non-negative</u>:

$$\delta \geqq 0 \ . \tag{27.22}$$

On the other hand, the <u>Clausius-Duhem inequality (27.13) holds and only if</u>

$$\begin{aligned} \frac{\underset{\sim}{h}\cdot\underset{\sim}{g}}{\theta} &\geqq -\delta \ , \\ &= \rho(\dot{\psi} + \eta\dot{\theta}) - w \ . \end{aligned} \tag{27.23}$$

We shall call (27.23) the reduced dissipation inequality, where the term "reduced" refers to the use of principles of balance of linear momentum, moment of momentum, and energy in order to derive this inequality from the assumption (27.13).

In summary, the two fundamental laws now subjoined to the laws of mechanics are the law of balance of energy (26.1) (with the specializing assumptions (27.1), (27.2), (27.4), and (27.5)) and the Clausius-Duhem dissipation inequality (27.12). We shall need to use them only through the reduced dissipation inequality (27.23). We shall not assume the truth of the Clausius-Planck inequality (27.22). Rather, we shall find that it can sometimes be proved to hold as a theorem.

Thermodynamic Processes.

Consider a set of functions consisting of, first, a motion $\underset{\sim}{x} = \underset{\sim}{\chi}(X,t)$ of a body $\mathcal{B}$, and, second, frame-indifferent fields $\theta(\underset{\sim}{x}, t)$, $\underset{\sim}{T}(\underset{\sim}{x}, t)$, $\underset{\sim}{h}(\underset{\sim}{x}, t)$, $\psi(\underset{\sim}{x}, t)$, and $\eta(\underset{\sim}{x}, t)$ defined over the configuration $\mathcal{B}_{\underset{\sim}{\chi}}$ of $\mathcal{B}$ at time t. These fields are said to constitute a thermodynamic process if they satisfy the principles of balance of momentum, moment of momentum, and energy, for every part of $\mathcal{B}$. By proper choice of body force $\underset{\sim}{b}$ and supply of heat working, s, the first and third principles, expressed in the forms (3.16) and (27.9) can always be satisfied, and the only restriction is imposed by Cauchy's second law, (3.18), viz, the stress tensor $\underset{\sim}{T}$ is symmetric: $\underset{\sim}{T} = \underset{\sim}{T}^T$.

Constitutive Equations. Principle of Dissipation.

Let the two histories χ^t, θ^t be called the deformation-temperature history. A thermodynamic constitutive equation is a rule whereby the deformation-temperature history determines unique stress, heat efflux, specific free energy, and specific entropy:

$$\left.\begin{matrix} \underset{\sim}{T} \\ \underset{\sim}{h} \\ \psi \\ \eta \end{matrix}\right\} \text{ are functionals of } \left\{\begin{matrix} \chi^t \\ \theta^t \end{matrix}\right. . \tag{27.24}$$

This theory generalizes one I proposed in 1949 and renders it more specific by use of recent ideas; it also extends COLEMAN's generalization of NOLL's principle of determinism (Lecture 3) to thermodynamic processes. It satisfies also the guiding principle for formulation of constitutive equations called the principle of equipresence: The independent variables in all constitutive equations of a theory shall be the same. In this case those variables are $\underset{\sim}{\chi}^t$ and θ^t.

The functionals entering a thermodynamic constitutive equation (27.24) are subject to three requirements:

1. The principle of local action.
2. The principle of material frame-indifference.
3. The principle of dissipation.

The first two of these are straightforward generalizations of their special cases stated in Lecture 3 for dynamic processes and need not be detailed here. The third is different. The principle of dissipation requires that the functionals in the constitutive equation (27.22) be such that the reduced dissipation inequality (27.21) holds for every thermodynamic process. That is, the Clausius-Duhem inequality is interpreted as an identical restriction upon constitutive equations. For the Navier-Stokes theory of fluids, it has been so interpreted for over fifty years, as witnessed by the DUHEM-STOKES inequalities (27.21). The idea of using it as an identical requirement for constitutive equations in general was first suggested and applied by NOLL.

We shall consider the consequences of the principle of dissipation in a major special case, that of the simple material.

Simple Thermodynamic Materials.

Let Φ stand for the collection $(\underset{\sim}{T}, \underset{\sim}{h}, \psi, \eta)$. A thermodynamic material is simple if (27.24) reduces to

$$\Phi = \mathfrak{S}(\underset{\sim}{F}^t, \theta^t, \underset{\sim}{g}^t), \qquad (27.25)$$

where, as before, $\underset{\sim}{F} \equiv \nabla\underset{\sim}{\chi}$, $\underset{\sim}{g} \equiv \operatorname{grad}\theta$. Clearly the simple materials of Lecture 3 are included as a special case, and clearly the principle of local action is satisfied.

Exercise 27.5. For the four constitutive equations represented schematically by (27.25), find reduced forms such as to satisfy the principle of material frame-indifference.

Quasi-elastic Response. Fading Memory.

Any history $f^t(s)$ for $s \geq 0$ may be regarded as consisting in its past history f^t_+, which is defined as $f^t(s)$ for $s > 0$, and its present value $f^t(0) = f(t)$. Therefore we may write (27.23) in the form

$$\Phi = \mathfrak{S}(\underset{\sim}{F}, \theta, \underset{\sim}{g}\ ;\ \underset{\sim}{F}^t_+,\ \theta^t_+,\ \underset{\sim}{g}^t_+)\ , \tag{27.26}$$

where $\mathfrak{S}$ is a functional of its last three arguments and a function of the first three. At any given instant t, the past histories at a particular particle have taken place, and according to (27.26) Φ depends upon the present values $\underset{\sim}{F}$, θ, $\underset{\sim}{g}$. That is,

$$\Phi = \mathfrak{f}(\underset{\sim}{F}, \theta, \underset{\sim}{g}, t) \tag{27.27}$$

where $\mathfrak{f}$ is a function. Now for a given functional $\mathfrak{S}$ in (27.26), i.e., for a given material, the function $\mathfrak{f}$ in (27.5) will depend upon what the past history is in (27.26). Thus $\mathfrak{f}$ does not express a material property alone, but rather the effect of a particular past history upon a certain material. The quantity Φ for two different particles in a homogeneous body will generally be given by two different $\mathfrak{f}$ in (27.27), though by the same $\mathfrak{S}$ in (27.26), because the past history differs from one particle to another.

If the four functions symbolized by $\mathfrak{f}$ in (27.27) are continuously differentiable with respect to all four arguments, the response of the material to the particular past deformation-temperature history is said to be quasi-elastic. The term arises from the fact that the dependence of $\underset{\sim}{T}$ on $\underset{\sim}{F}$ is just like that of an elastic material (11.2) except that the response is time-dependent. Here, as we shall see shortly, it is the smoothness of the four functions denoted collectively by $\mathfrak{f}$ that makes all the difference.

If $\mathfrak{f}$ is continuously differentiable with respect to t, a small change in the past history effects an approximately linear change in Φ, and likewise for changes in the present values $\underset{\sim}{F}$, θ, and $\underset{\sim}{g}$. This smoothness in itself is an assertion of fading memory in one sense. In a difficult analysis COLEMAN has shown that if $\mathfrak{C}$ satisfies a generalized principle of fading memory in COLEMAN & NOLL's sense, the response of the material is quasi-elastic. This theorem is far too difficult for us to prove here. As was shown by BOWEN & WANG, quasi-elastic response by itself, for whatever cause, suffices to yield a formal theory similar to the thermodynamics of homogeneous processes.

Before showing how this is so, we remark that the Navier-Stokes fluid does not exhibit quasi-elastic response. For it, if we hold $\underset{\sim}{F}$, θ, and $\underset{\sim}{g}$ fixed but change t, we do not get even a small change in Φ itself, since $\underset{\sim}{T}$ is determined by $\dot{\underset{\sim}{F}}\underset{\sim}{F}^{-1}$. Likewise, more general Rivlin-Ericksen materials do not exhibit quasi-elastic response. Thus quasi-elastic response is not a general property of materials but rather a distinguishing attribute of an important special class.

Dissipation and Quasi-elastic Response.

We now write (27.27) explicitly:

$$\begin{aligned} \underset{\sim}{T} &= \underset{\sim}{\mathfrak{g}}(\underset{\sim}{F}, \theta, \underset{\sim}{g}, t) , \\ \underset{\sim}{h} &= \underset{\sim}{\mathfrak{h}}(\underset{\sim}{F}, \theta, \underset{\sim}{g}, t) , \\ \psi &= \mathfrak{y}(\underset{\sim}{F}, \theta, \underset{\sim}{g}, t) , \\ \eta &= \mathfrak{h}(\underset{\sim}{F}, \theta, \underset{\sim}{g}, t) , \end{aligned} \tag{27.28}$$

and substitute these formulae into the reduced dissipation inequality (27.23), obtaining

$$\begin{aligned} -(\partial_{\underset{\sim}{g}}\mathfrak{y})\dot{\underset{\sim}{g}} - \rho(\partial_\theta\mathfrak{y} + \eta)\dot{\theta} + \operatorname{tr}\{[\underset{\sim}{F}^{-1}\underset{\sim}{T} - \rho(\partial_{\underset{\sim}{F}}\mathfrak{y})^T]\dot{\underset{\sim}{F}}\} \\ + \frac{1}{\theta}\underset{\sim}{h}\cdot\underset{\sim}{g} - \rho\partial_t\mathfrak{y} \geqq 0 . \end{aligned} \tag{27.29}$$

With any values we please for $\underset{\sim}{F}$, θ, and $\underset{\sim}{g}$, we can find a process such that at a given place and time $\dot{\underset{\sim}{g}} = \underset{\sim}{0}$, $\dot{\theta} = 0$, $\dot{\underset{\sim}{F}} = \underset{\sim}{0}$. Hence (27.29) yields the necessary condition

$$\frac{1}{\theta}\,\mathbf{h}\cdot\mathbf{g} - \rho\partial_t \mathfrak{y} \geqq 0\ . \tag{27.30}$$

The left-hand side is a function of $\mathbf{F}$, θ, $\mathbf{g}$, and t alone, by (27.28). Giving $\mathbf{F}$, θ, $\mathbf{g}$ any values we please, we can find a process in which $\dot{\mathbf{F}} = \mathbf{0}$, $\dot{\theta} = 0$, and $\dot{\mathbf{g}}$ has any magnitude and direction we please. Hence the first term in (27.29) may be given a value such as to violate (27.30) unless $\partial_{\mathbf{g}}\mathfrak{y} = \mathbf{0}$. That is, the free energy does not depend on the temperature gradient:

$$\psi = \mathfrak{y}\,(\mathbf{F},\ \theta,\ t)\ . \tag{27.31}$$

Similar reasoning shows that the next two terms in (27.29) vanish:

$$\eta = -\,\partial_\theta \mathfrak{y}\ , \tag{27.32}$$

and

$$\operatorname{tr}\{[\mathbf{F}^{-1}\mathbf{T} - \rho(\partial_{\mathbf{F}}\mathfrak{y})^T]\dot{\mathbf{F}}\} = 0\ . \tag{27.33}$$

Exercise 27.6. If $\mathfrak{y}$ is a frame-indifferent scalar, prove that $\mathbf{F}(\partial_{\mathbf{F}}\mathfrak{y})^T$ is a symmetric tensor, and hence that (27.33) requires

$$\mathbf{T} = \rho\mathbf{F}(\partial_{\mathbf{F}}\mathfrak{y})^T = \rho(\partial_{\mathbf{F}}\mathfrak{y})\mathbf{F}^T\ . \tag{27.34}$$

Hence

$$w = \rho\ \operatorname{tr}(\partial_{\mathbf{F}}\mathfrak{y})\dot{\mathbf{F}}^T\ . \tag{27.35}$$

These results state that the free-energy function $\mathfrak{y}\,(\mathbf{F},\ \theta,\ t)$ is a thermodynamic potential for the specific entropy η and the stress $\mathbf{T}$. That is, the effects of memory upon the free energy determine all effects of memory on η and $\mathbf{T}$. Moreover, comparison with (20.1) shows that $\mathfrak{y}$ plays the role of a stored-energy function in hyperelasticity.

We have not finished with the consequences of the reduced dissipation inequality. If we substitute (27.35), (27.32), and (27.31) into the definition (27.17) of the internal dissipation, we obtain

$$\delta = -\,\rho\partial_t \mathfrak{y}\ . \tag{27.36}$$

That is, the internal dissipation is the rate of decrease of the free energy per unit volume at constant deformation and temperature. Since we interpret the time dependence of quasi-elastic response as an overall effect of material memory, the result (27.36) may be regarded as stating, loosely, that δ is the rate of decrease of free energy per unit volume due to effects of memory alone. Naturally, we do not expect the material to gain free energy as a result of remembering the past; that is, we expect this decrease to be non-positive, and shortly we shall prove that it is so indeed.

Namely, we look back at our standing assumption (27.23) and observe that by (27.35) and (27.31), the right-hand side is independent of $\underset{\sim}{g}$. Hence the left-hand side is an upper bound for $-\delta$ for every $\underset{\sim}{g}$, and in particular when $\underset{\sim}{g} = \underset{\sim}{0}$. Hence

$$\delta \geqq 0 \quad . \tag{27.37}$$

That is, the internal dissipation is non-negative. In other words, for quasi-elastic response the Clausius-Planck inequality is a proved theorem if the Clausius-Duhem inequality is assumed.

<u>Exercise 27.7.</u> Prove that, conversely, if

1. $\mathfrak{Y}(\underset{\sim}{F}, \theta, t)$ is a frame-indifferent scalar function,
2. $\partial_t \mathfrak{Y} \leq 0$,
3. η and $\underset{\sim}{T}$ are defined by (27.32) and (27.34),
4. $\underset{\sim}{h}$ satisfies (27.23) with δ defined by (27.17) or, alternatively, by (27.36),

the equations (27.26) of quasi-elastic response satisfy the reduced dissipation inequality (27.23) in every thermodynamic process.

The foregoing results constitute the <u>grand thermodynamic theorem</u> of COLEMAN, as generalized by WANG & BOWEN. That theorem states, roughly, that any material with quasi-elastic response always consistent with the Clausius-Duhem dissipation inequality obeys essentially the equations of the classical, special thermodynamics of homogeneous processes, based on caloric equations of state (Lecture 26, especially (26.11) and (26.13)), provided that these equations relate the densities of the state variables and be time-dependent. In addition, the internal dissipation is non-negative, and the classical heat-conduction inequality is replaced by one giving a lower bound for $\underset{\sim}{h} \cdot \underset{\sim}{g}$ in terms of the internal dissipation. If the product

$\theta\delta$ is large enough, indicating that there is sufficiently great internal dissipation at sufficiently high temperature, the heat efflux vector $\underset{\sim}{h}$ may point even in the opposite direction from $\underset{\sim}{g}$, indicating that heat may be drawn straight from a cold part to a hot part, the opposite of the behavior we have been led to expect by experience with materials in which $\delta = 0$.

Applications:

To develop alternative forms and applications of COLEMAN's theorem would require a course of lectures. I remark here on only a few. First, under suitable hypotheses of inversion, the functions $\varepsilon(\underset{\sim}{F}, \eta, t)$ may be used as a potential. Second, the fact that $\delta \geqq 0$ makes it possible to prove, rigorously and without further assumptions, explicit theorems on minimum free energy, maximum entropy, and minimum internal energy, and a general theorem on cyclic processes, corresponding to the rather vague assertions of this kind in thermodynamics books. Third, if we construct a sequence of deformation-temperature histories $\underset{\sim}{F}^t_\alpha$, θ^t_α, such that $\underset{\sim}{F}^t_\alpha$, and $\mathfrak{y}(\underset{\sim}{F}^t_\alpha, \theta^t_\alpha, t)$ approach time-independent limits $\underset{\sim}{F}_o$, θ_o, and $\psi(\underset{\sim}{F}_o, \theta_o)$ as $\alpha \to 0$, we can expect that the equilibrium thermodynamics is corresponding to a free-energy function $\psi(\underset{\sim}{F}_o, \theta_o)$ should result. COLEMAN has constructed two such sequences, the former corresponding to retardation (Lecture 24), and the latter to an infinitely fast process, and has proved that indeed classical thermodynamics based on caloric equations of state emerges in both cases. The limit functions $\psi(\underset{\sim}{F}_o, \theta_o)$ are of course different, determined by the limit process and the nature of the original functional $\mathfrak{y}(\underset{\sim}{F}^t, \theta^t, t)$. That is, two <u>different</u> quasi-elastic (in fact elastic) responses result in the two limit cases. Thus the theory based on local caloric equations of state emerges in four ways:

1. As a theory of certain special simple materials, for all processes.
2. As the equilibrium theory for all simple materials with quasi-elastic response.
3. In the limit of retardation.
4. In the limit of fast processes.

In the last two cases, certain assumptions of smoothness beyond that of quasi-elastic response are needed.

A fourth major application is to the theory of wave motions in dissipative materials, as we shall see in the next lecture.

A Final Remark on the Stored-Energy Function in Elasticity.

It is commonly believed that thermodynamics proves every elastic material to be hyperelastic. In order to apply thermodynamic reasoning, we must introduce heat and temperature, which are never mentioned in elasticity. Thus to get results from thermodynamics we must first accept more assumptions than are necessary to obtain the theory of elasticity. A fortiori, thermodynamics cannot prove anything about elasticity itself, though it can and does deliver results about thermoelasticity formulated within a thermodynamic framework:

$$\begin{aligned} \underset{\sim}{T} &= \underset{\sim}{\mathfrak{g}}(\underset{\sim}{F}, \theta) , \\ \psi &= \psi(\underset{\sim}{F}, \theta) , \\ \eta &= \eta(\underset{\sim}{F}, \theta) , \\ \underset{\sim}{h} &= \underset{\sim}{\mathfrak{h}}(\underset{\sim}{F}, \theta, \underset{\sim}{g}) . \end{aligned} \tag{27.38}$$

Clearly, in the terms used before in these lectures, the thermoelastic material so defined is not even an elastic material, in general, since $\underset{\sim}{T}$ depends on θ as well as $\underset{\sim}{F}$. The theorems on quasi-elastic response may be applied. They yield the following results: The constitutive equations of the thermoelastic material (27.38) are consistent with the reduced dissipation inequality (27.23) for all thermodynamic processes if and only if they reduce to

$$\begin{aligned} \psi &= \psi(\underset{\sim}{F}, \theta) , \\ \underset{\sim}{T}_{\kappa} &= \rho_{\kappa} \partial_{\underset{\sim}{F}} \psi = \rho_{\kappa} \partial_{\underset{\sim}{F}} \varepsilon , \\ \eta &= - \partial_{\theta} \psi , \qquad \theta = \partial_{\eta} \varepsilon , \\ \underset{\sim}{h} \cdot \underset{\sim}{g} &\geqq 0 , \end{aligned} \tag{27.39}$$

where $\varepsilon(\underset{\sim}{F}, \eta) \equiv \psi(\underset{\sim}{F}, \mathfrak{t}(\underset{\sim}{F}, \eta)) + \eta\, \mathfrak{t}(\underset{\sim}{F}, \eta)$ and $\mathfrak{t}(\underset{\sim}{F}, \eta)$ is the function obtained by solving $(27.38)_3$ for θ. In particular, the thermoelastic material is a hyperelastic material for two special cases:

1. In isothermal deformations, with $\sigma(\underset{\sim}{F}) = \psi(\underset{\sim}{F}, \theta)$ and $\theta = \text{const.}$
2. In isentropic deformation, with $\sigma(\underset{\sim}{F}) = \varepsilon(\underset{\sim}{F}, \eta)$ and $\eta = \text{const.}$

The very fact that the potential relations $(27.39)_{2,3}$ hold in <u>all</u> deformations of the thermoelastic material shows that in general, when θ or η varies from point to point, the thermoelastic material is <u>not elastic</u> and hence, <u>a fortiori</u>, cannot have a stored-energy function.

<u>Exercise 27.8</u>. Find analogues of (27.39) when η is given as a function of $\underset{\sim}{F}$ and ε. Show that a thermoelastic material is generally not hyperelastic in iso-energetic deformations.

LECTURE 28: WAVE PROPAGATION IN DISSIPATIVE MATERIALS.

The "Smoothing" Effect of Dissipation.

There is a widespread belief that dissipation "smoothes out" discontinuities. The source of this belief seems to lie in a fact about linearly viscous fluids. Indeed, consider the Navier-Stokes formula (23.17), *viz*

$$\underset{\sim}{T} = - p(\rho)\underset{\sim}{1} + \lambda(\mathrm{tr}\underset{\sim}{D})\underset{\sim}{1} + 2\mu\underset{\sim}{D} \quad . \tag{23.17}$$

By (18.8) and (18.9), the geometrical and kinematical conditions at an acceleration wave with normal $\underset{\sim}{n}$ imply that

$$[\mathrm{tr}\underset{\sim}{D}] = - U\, \underset{\sim}{a}\cdot\underset{\sim}{n} \quad , \quad [2\underset{\sim}{D}] = - U(\underset{\sim}{a} \otimes \underset{\sim}{n} + \underset{\sim}{n} \otimes \underset{\sim}{a}) \quad , \tag{28.1}$$

where $\underset{\sim}{a}$ is the amplitude and U is the local speed of propagation. Hence

$$\begin{aligned} [\underset{\sim}{T}] &= - U\{\lambda(\underset{\sim}{a}\cdot\underset{\sim}{n})\underset{\sim}{1} + \mu(\underset{\sim}{a} \otimes \underset{\sim}{n} + \underset{\sim}{n} \otimes \underset{\sim}{a}) \quad , \\ \underset{\sim}{0} = [\underset{\sim}{T}]\underset{\sim}{n} &= - U\{(\lambda + \mu)(\underset{\sim}{a}\cdot\underset{\sim}{n})\underset{\sim}{n} + \mu\underset{\sim}{a}\} \quad , \end{aligned} \tag{28.2}$$

where the vanishing of this last expression is a consequence of Poisson's condition (18.11). If $U = 0$, there is no wave. If $U \neq 0$, taking the scalar products of (28.2) by $\underset{\sim}{n}$ and $\underset{\sim}{a}$ implies that

$$\begin{aligned} (\lambda + 2\mu)(\underset{\sim}{a}\cdot\underset{\sim}{n}) &= 0 \quad , \\ (\lambda + \mu)(\underset{\sim}{a}\cdot\underset{\sim}{n})^2 + \mu a^2 &= 0 \quad . \end{aligned} \tag{28.3}$$

By the Duhem-Stokes inequalities (27.21), $\lambda + 2\mu > 0$, unless $\mu = 0$, and hence (28.3) requires that in a viscous fluid $a = 0$. That is, the Navier-Stokes constitutive equation is incompatible with the existence of acceleration waves unless $\lambda = \mu = 0$.

True though this fact is, it does not justify any sweeping conclusions about the effects of "dissipative mechanisms" on surfaces of discontinuity. For example, the partial differential equation governing the infinitesimal displacement u of a vibrating string subject to linear viscosity is

$$\sigma\partial_t^2 u = T\partial_x^2 u - K\partial_t u \quad , \tag{28.4}$$

where σ is the line density, T is the tension, and K is a positive constant. Certainly a "dissipative mechanism" is represented here, yet (28.4) admits discontinuous solutions of all orders, and all have the classical speed of propagation given by $U^2 = T/\sigma$, no matter what be the value of K.

Though in both these examples the dissipative mechanism is called "linear friction" by the physicists, clearly such different results must correspond to ideas that in any true physics are quite different from one another. We have seen in Lecture 22 that the Navier-Stokes fluid does not have fading memory in COLEMAN & NOLL's sense, and in the last lecture we have seen that it does not show quasi-elastic response. Neither of these statements can be made about (28.4), which does not fall into any of the types of constitutive equations we have studied.

As we shall see now, quasi-elastic response is generally compatible with the propagation of weak waves.

Waves and Quasi-Elastic Response.

The theory of quasi-elastic response from the previous lecture and the theory of wave propagation in elastic materials, presented in Lecture 18, may be combined. The results are due in principle to COLEMAN & GURTIN, but the simple approach I shall follow here was first outlined by WANG & BOWEN.

First, although in all the analysis in the preceding lecture a single particle was considered, we must now take account of the variation of material properties from one particle to another. Even in a homogeneous material with fading memory, the quasi-elastic response function $\mathfrak{y}$ giving the specific free energy will generally vary from one particle to another because the deformation-temperature history will not generally be homogeneous. Thus (27.31) is replaced by

$$\psi = \mathfrak{h}(\underset{\sim}{F}, \theta, t, \underset{\sim}{X}) . \tag{28.5}$$

By (11.13) and (27.34), the Piola-Kirchhoff stress $\underset{\sim}{T}_{\underset{\sim}{\kappa}}$ is given by

$$\underset{\sim}{T}_{\underset{\sim}{\kappa}} = \rho_{\underset{\sim}{\kappa}} \partial_{\underset{\sim}{F}} \mathfrak{h} . \tag{28.6}$$

In the definition of quasi-elastic response, <u>both</u> ψ and $\underset{\sim}{T}$ were required to be continuously differentiable functions of $\underset{\sim}{F}$, θ, and t. By the result (28.6), also $\partial_{\underset{\sim}{F}} \mathfrak{h}$ is shown to be continuously differentiable in these same variables. We shall assume also that the quasi-elastic response varies smoothly from one particle to the next in the sense that $\mathfrak{h}$ is continuously differentiable with respect to $\underset{\sim}{X}$.

Cauchy's first law of motion assumes the explicit form (11.14), <u>viz</u>

$$\text{Div}\, \underset{\sim}{T}_{\underset{\sim}{\kappa}} + \rho_{\underset{\sim}{\kappa}} \underset{\sim}{b} = \rho_{\underset{\sim}{\kappa}} \ddot{\underset{\sim}{x}} . \tag{11.14}$$

By (28.5) and (28.6), the tensor $\underset{\sim}{T}_{\underset{\sim}{\kappa}}$ now depends on θ and t as well as $\underset{\sim}{F}$ and $\underset{\sim}{X}$. To spare indices, we shall use a schematic direct notation. Setting

$$\tilde{A} \equiv \rho_{\underset{\sim}{\kappa}} \partial^2_{\underset{\sim}{F}} \mathfrak{h} ,$$
$$\tilde{\underset{\sim}{q}} \equiv \partial_{\underset{\sim}{X}} \cdot (\rho_{\underset{\sim}{\kappa}} \partial_{\underset{\sim}{F}} \mathfrak{h}) , \tag{28.7}$$

from (11.14) we have

$$\text{tr}(\tilde{A}[\nabla \underset{\sim}{F}]) + \rho_{\underset{\sim}{\kappa}} \text{tr}[(\partial_\theta \partial_{\underset{\sim}{F}} \mathfrak{h}) \nabla \theta] + \tilde{\underset{\sim}{q}} + \rho_{\underset{\sim}{\kappa}} \underset{\sim}{b} = \rho_{\underset{\sim}{\kappa}} \ddot{\underset{\sim}{x}} . \tag{28.8}$$

Now we calculate the jump of this equation at an acceleration wave. As I mentioned above, even in a homogeneous material with quasi-elastic response we shall generally have $\tilde{\underset{\sim}{q}} \neq \underset{\sim}{0}$. However, by $(28.7)_2$ and our assumptions, $\tilde{\underset{\sim}{q}}$ is a continuous function of $\underset{\sim}{F}$, θ, and $\underset{\sim}{X}$. By the definition of an acceleration wave, $\underset{\sim}{F}$ is continuous across it. We shall add the assumption that θ is also continuous. That is, the wave is not a thermal shock. Then

$$[\tilde{\mathbf{q}}] = \mathbf{0} \quad , \tag{28.9}$$

even though $\tilde{\mathbf{q}} \neq \mathbf{0}$. We shall add the assumption that the temperature gradient suffers no discontinuity at the wave:

$$[\nabla\theta] = \mathbf{0} \quad . \tag{28.10}$$

Such a wave is called homothermal, because it is most easily visualized as travelling through a region of uniform temperature. For a homothermal acceleration wave, taking the jump of (28.8) yields a propagation condition of just the same form as (18.17), viz

$$(\tilde{\mathbf{Q}}(\mathbf{n}) - \rho\tilde{U}^2\mathbf{1})\mathbf{a} = \mathbf{0} \quad , \tag{28.11}$$

where the tildes indicate that the elasticity is $\tilde{A}$, calculated from $\mathfrak{y}(\mathbf{F}, \theta, t, \mathbf{X})$ by $28.7)_1$. Hence, corresponding to the result for elasticity in Lecture 20, $\tilde{\mathbf{Q}}(\mathbf{n})$ is symmetric:

$$\tilde{\mathbf{Q}}(\mathbf{n}) = \tilde{\mathbf{Q}}(\mathbf{n})^T \quad . \tag{28.12}$$

The results so far may be summarized as follows: the propagation condition for homothermal waves in a material with quasi-elastic response is that of a hyperelastic material with time-dependent stored-energy function. The corresponding acoustical tensor is calculated from the isothermal elasticity $\tilde{A}$.

Thus the dissipative mechanisms embodied in quasi-elastic response subject to the Clausius-Duhem dissipation inequality lead to constitutive equations in no way incompatible with wave motion. Quite the reverse, the laws of wave propagation are at each place and time as that of some hyperelastic material. In the interpretation for materials with fading memory, we may say that the past history of temperature and deformation as well as their present values determine the acoustic tensor.

As mentioned in the preceding lecture, we can just as well use the specific internal-energy function $\mathfrak{e}(F, \eta, t, \mathbf{X})$ as a thermodynamic potential. Let quantities calculated from this potential be denoted by a caret: $\hat{A}$, $\hat{\mathbf{Q}}(\mathbf{n})$, etc. Parallel reasoning, based on the assumptions that not only η but also $\nabla\eta$ is continuous,

$$[\nabla\eta] = \underset{\sim}{0} \quad , \tag{28.13}$$

leads to a result of the same form as (28.11) with carets replacing tildes:

$$(\hat{\underset{\sim}{Q}}(\underset{\sim}{n}) - \rho\hat{U}^2\underset{\sim}{1})\hat{\underset{\sim}{a}} = \underset{\sim}{0} \quad . \tag{28.14}$$

Waves satisfying (28.13) are called <u>homentropic</u>, and (28.14) is the propagation condition for them.

In general, waves will not be both homentropic and homothermal. The results (28.11) and (28.14) refer to <u>different</u> waves, and conceivably there are still other kinds in which neither assumption is valid. Before determining circumstances in which the results apply, we shall find a relation between the two acoustic tensors $\hat{\underset{\sim}{Q}}$ and $\tilde{\underset{\sim}{Q}}$

<u>Relation Between the Homothermal and Homentropic Acoustic Tensors.</u>

By the results in the last lecture, the Piola-Kirchhoff stress is given not only by (28.6) but also by its counterpart when the internal energy function $\mathfrak{e}$ is used as a potential:

$$\underset{\sim}{T}_{\kappa} = \rho_{\kappa}\partial_{\underset{\sim}{F}}\mathfrak{y}(\underset{\sim}{F}, \theta, t) = \rho_{\kappa}\partial_{\underset{\sim}{F}}\mathfrak{e}(\underset{\sim}{F}, \eta, t) \ . \tag{28.15}$$

These equations are valid in all motions of the material with quasi-elastic response subject to the Clausius-Duhem inequality. We are assuming further that the relation (27.32), <u>viz</u> $\eta = -\partial_{\theta}\mathfrak{y}(\underset{\sim}{F}, \theta, t)$, may be solved for θ:

$$\theta = \mathfrak{t}(\underset{\sim}{F}, \eta, t) \ , \tag{28.16}$$

and that $\mathfrak{t}$ is continuously differentiable. If we differentiate (28.15) with respect to $\underset{\sim}{F}$ and use the chain rule, we obtain by $(28.7)_1$ and its counterpart for $\mathfrak{e}$

$$\begin{aligned} \frac{1}{\rho_{\kappa}}(\hat{A} - \tilde{A}) &= \mathrm{tr}[(\partial_{\theta}\partial_{\underset{\sim}{F}}\mathfrak{y}) \otimes (\partial_{\eta}\partial_{\underset{\sim}{F}}\mathfrak{y}(\underset{\sim}{F}, \mathfrak{t}(\underset{\sim}{F}, \eta), t))] \\ &= \mathrm{tr}[(\partial_{\theta}\partial_{\underset{\sim}{F}}\mathfrak{y}) \otimes (\partial_{\theta}\partial_{\underset{\sim}{F}}\mathfrak{y})(\partial_{\eta}\mathfrak{t})] \ . \end{aligned} \tag{28.17}$$

Forming the acoustic tensors $\hat{\underset{\sim}{Q}}$ and $\tilde{\underset{\sim}{Q}}$ by the counterparts of (18.16) yields

$$\hat{\mathbf{Q}} - \tilde{\mathbf{Q}} = \rho(\partial_\eta \mathfrak{t})\,\mathrm{tr}[(\partial_\theta \partial_{\mathbf{F}} \mathfrak{h})\mathbf{F}^T \mathbf{n} \otimes (\partial_\theta \partial_{\mathbf{F}} \mathfrak{h})\mathbf{F}^T \mathbf{n}] \quad . \tag{28.18}$$

Now if we write (27.32) in the form $\eta = \mathfrak{h}(\mathbf{F}, \theta, t)$ and define the specific heat at constant deformation $\kappa_{(\mathbf{F})}$

$$\kappa_{(\mathbf{F})} \equiv \theta \partial_\theta \mathfrak{h} \, , \tag{28.19}$$

we have

$$\partial_\eta \mathfrak{t}(\mathbf{F}, \eta) = \frac{1}{\partial_\theta \mathfrak{h}(\mathbf{F}, \theta)} = \frac{\theta}{\kappa_{(\mathbf{F})}} \tag{28.20}$$

Therefore if we set

$$\mathbf{p}(\mathbf{n}) \equiv \sqrt{\frac{\rho\theta}{\kappa_{(\mathbf{F})}}}\,\mathrm{tr}\{(\partial_\theta \partial_{\mathbf{F}} \mathfrak{h})\mathbf{F}^T \mathbf{n}\} \, , \tag{28.21}$$

we obtain <u>DUHEM's identity</u> connecting the isothermal and isentropic acoustic tensors:

$$\hat{\mathbf{Q}} = \tilde{\mathbf{Q}} + \mathbf{p} \otimes \mathbf{p} \quad . \tag{28.22}$$

The quadratic form of $\mathbf{p} \otimes \mathbf{p}$ is non-negative definite. Hence if $\tilde{\mathbf{Q}}$ is positive definite, so is $\hat{\mathbf{Q}}$. This result, too, is due to DUHEM. In the terms used in Lecture 18, we may express it as follows: If the isothermal elasticity $\tilde{A}$ is strongly elliptic, so is the isentropic elasticity $\hat{A}$. Since $\tilde{\mathbf{Q}}(\mathbf{n})$ is symmetric, it has real principal axes; therefore, if the proper number of $\tilde{\mathbf{Q}}(\mathbf{n})$ are non-negative, then $\tilde{\mathbf{Q}}(\mathbf{n})$ is non-negative definite. By (28.22), so is $\hat{\mathbf{Q}}(\mathbf{n})$. Thus follows another form of DUHEM's theorem: <u>If for a given direction of propagation the isothermal wave speeds are real, so are the isentropic ones</u>. The special case of this theorem in gas dynamics in familiar, since the two squared speeds are $\partial_\rho \tilde{p}(\rho,\theta)$ and $\partial_\rho \hat{p}(\rho,\eta)$, and the ratio of these quantities is the positive constant denoted usually by γ.

<u>Waves in Non-Conductors</u>.

To obtain $\hat{\mathbf{Q}}$ as the acoustic tensor, we assumed that both η and $\nabla\eta$ are continuous across the wave, and we called such waves homentropic. In some materials, as we shall see now, all acceleration waves are necessarily homentropic. We shall lay down the conditions

$$[\eta] = 0 \quad , \quad [\theta] = 0 \quad , \tag{28.23}$$

as part of the definition of a second-order singular surface. That is, we assume that the singularity is not a shock with respect to entropy or temperature as well as deformation. For materials with quasi-elastic response, the former condition follows from the latter because of the results (27.31) and (27.32) of COLEMAN's theorem; for materials having caloric equations of state in the ordinary sense, it follows directly.

We consider <u>non-conductors of heat</u>, namely, those materials for which the constitutive equation $(27.28)_2$ takes the form

$$\underset{\sim}{h} \equiv \underset{\sim}{0} \quad . \tag{28.24}$$

(In fact, for the argument I shall give it suffices that

$$[\operatorname{div} \underset{\sim}{h}] = 0 \quad , \tag{28.25}$$

but this condition does not seem to correspond to any natural model except under the stronger assumption in (28.24).) By (28.23) and MAXWELL's theorem (18.5), there exists a scalar B such that

$$[\operatorname{grad} \eta] = B\underset{\sim}{n} \quad , \tag{28.26}$$

and by HADAMARD's kinematical condition

$$[\dot{\eta}] = -\, UB \ . \tag{28.27}$$

By (27.18) and the result (27.36) of COLEMAN's theorem, in quasi-elastic response

$$-\,\rho\partial_t \mathfrak{y} = \rho\theta\dot{\eta} - \operatorname{div} \underset{\sim}{h} - \rho s \ . \tag{28.28}$$

Moreover, the left-hand side is a function of ρ, θ, $\underset{\sim}{F}$, and t alone, by (27.31). Taking the jump of (28.28) yields by (28.25) and (28.27)

$$0 = \rho\theta[\dot{\eta}] = -\,\rho\theta UB \quad . \tag{28.29}$$

If the discontinuity is indeed a wave, $U \neq 0$, so $B = 0$, so that $[\text{grad}\ \eta] = \mathbf{0}$ by (28.26). Thus we have the following theorem: Every acceleration wave in a non-conductor with quasi-elastic response is homentropic. This result, due in stages to DUHEM and to COLEMAN & GURTIN, includes and generalizes a classic result of gas dynamics: when heat conduction is neglected, sound waves are isentropic, and their squared speed is $\partial_\rho \hat{p}(\rho,\eta)$.

Waves in Definite Conductors.

To obtain $\tilde{\mathbf{Q}}$ as the acoustic tensor, we assumed that both θ and $\nabla\theta$ are continuous across the wave, and we called such waves homothermal. In some materials, as we shall see now, all acceleration waves are necessarily homothermal. Naturally these materials must exclude the non-conductors, just considered.

Exercise 28.1. By applying KOTCHINE's theorem (18.10) to (26.1) with Q given by (27.6), show that the energy balance at a weak singularity at which momentum is balanced is equivalent to FOURIER's condition:

$$[\mathbf{h}] \cdot \mathbf{n} = 0 \quad . \tag{28.30}$$

That is, the jump of the heat efflux vector must be transverse.

We consider the constitutive equation $(27.26)_2$ for the heat flux. Defining the conductivity by

$$\mathbf{K} \equiv \partial_{\mathbf{g}} \boldsymbol{\mathfrak{h}}(\mathbf{F},\ \theta,\ \mathbf{g},\ t) \ , \tag{28.31}$$

we notice that in FOURIER's theory of heat conduction $\mathbf{h} = \boldsymbol{\mathfrak{h}}(\mathbf{F},\ \theta,\ \mathbf{g},\ t) = \mathbf{K}\mathbf{g}$, so that $\mathbf{K}$ reduces then to what is ordinarily called the "thermal conductivity." If the quadratic form of $\mathbf{K}$ is positive-definite,

$$\mathbf{m} \cdot \mathbf{K}\mathbf{m} > 0 \tag{28.32}$$

for every non-zero vector $\mathbf{m}$, we shall say that the material is a definite conductor of heat. (In the special case of Fourier's theory, a definite conductor is a material satisfying the classical heat-conduction inequality, but for more general materials no such interpretation of (28.32) suggests itself.) From (28.30) and $(27.28)_2$,

$$(\underset{\sim}{\mathfrak{h}}(\underset{\sim}{F}, \theta, \underset{\sim}{g} + [\underset{\sim}{g}], t) - \underset{\sim}{\mathfrak{h}}(\underset{\sim}{F}, \theta, \underset{\sim}{g}, t)) \cdot \underset{\sim}{n} = 0 \quad . \qquad (28.33)$$

At an acceleration wave, again (28.23) holds, so by MAXWELL's theorem there exists a scalar B such that

$$[\underset{\sim}{g}] = B\underset{\sim}{n} \quad . \qquad (28.34)$$

Therefore (28.33) may be written as

$$(\underset{\sim}{\mathfrak{h}}(\underset{\sim}{F}, \theta, \underset{\sim}{g} + B\underset{\sim}{n}, t) - \underset{\sim}{\mathfrak{h}}(\underset{\sim}{F}, \theta, \underset{\sim}{g}, t)) \cdot \underset{\sim}{n} = 0 \, . \qquad (28.35)$$

But if $B \neq 0$,

$$\underset{\sim}{n} \cdot d_B \underset{\sim}{\mathfrak{h}}(\underset{\sim}{F}, \theta, \underset{\sim}{g} + B\underset{\sim}{n}, t) = \underset{\sim}{n} \cdot \partial_{\underset{\sim}{g}} \underset{\sim}{\mathfrak{h}}(\underset{\sim}{F}, \theta, \underset{\sim}{g} + B\underset{\sim}{n}, t)\underset{\sim}{n} \, . \qquad (28.36)$$

By (28.35), the left-hand side vanishes. For a definite conductor, the right-hand side is positive. Hence $B = 0$. By (28.34), $[\underset{\sim}{g}] = \underset{\sim}{0}$. We have shown, then, that <u>every acceleration wave in a definite conductor of heat with quasi-elastic response is homothermal</u>. This theorem derives ultimately from FOURIER; in the generality given here, it is due to COLEMAN & GURTIN.

<u>Closure.</u>

Taken together, these two theorems show how important is heat conduction in determining the nature of wave motion. In general, of course, there is no reason to expect a material with quasi-elastic response to be either a non-conductor or a definite conductor, and hence we cannot justly regard either of the two foregoing theorems as indicating the kind of waves that do occur according to the general theory. The acoustic tensors $\hat{\underset{\sim}{Q}}$ and $\tilde{\underset{\sim}{Q}}$ have been shown to pertain to homentropic and homothermal waves, but acceleration waves in a material with quasi-elastic response cannot generally be expected to be of these kinds.

PART VI. STATICS

LECTURE 29: STABILITY IN ENERGY. VARIATIONAL THEOREMS.

The Independence of Statics.

In Lecture 11 we have seen that static elasticity is the general theory of statics of simple materials, and in Lectures 18-21 we have proved theorems which make it more or less reasonable to study especially the class of elastic materials which have stress potentials or stored-energy functions $\sigma(\underset{\sim}{F})$. In this approach, equilibrium is found as a special case of motion, and the existence of a potential follows from further assumptions about virtual motions.

It is always possible to formulate a theory of equilibrium in _purely static terms_ never mentioning motion or force, and, instead of laying down a constitutive equation at the start, to deduce its existence and form from other ideas. Of course, no theorem is proved from nothing; something must be assumed, and if the results come out to be the same in the end, the assumptions are equivalent. However, it may be enlightening to connect what seem at first different and independent trains of thought. Moreover, a good statics is never fully equivalent to a special case of motion, since it always carries with it some concept of _stability of equilibrium_.

Statics as an independent science rests on some kind of _minimal principle_. Of course I do not mean merely what the physicists call a "variational principle," where some formal operations with δ's are introduced so as to get a statement that looks a little more mysterious than an equivalent one directly motivated in the phenomena. Rather, I mean the assertion of a true minimum, carrying with it not only the vanishing of the first variation but also various inequalities. The difference is easily made plain in the familiar case of the catenary.

The Catenary as an Example.

The catenary is a particular curve. It is defined dynamically as follows. First, a general theory of one-dimensional continua is laid down, if often in the older treatments only by implication. The action of one part of the curve on its neighbor is postulated to be equipollent to that of a contact force and contact moment located at the junction of the two parts. As general laws of mechanics, equilibrium of forces and moments are laid down. A perfectly flexible line is then defined by a constitutive equation: The contact force is tangent to the curve, and the contact moment is zero. Combined, the constitutive equation and the laws of mechanics yield a differential equation. If the line density is uniform and the body force is a constant vector, an integral of this differential equation such as to pass through two assigned points and to have assigned length is easily found. In fact, there are generally two such integrals. The one curved upward is called "the perfect arch," while that curved downward is called "the catenary curve."

Exercise 29.1. Verify the foregoing outline by carrying out the steps. Show that for a perfectly flexible line, equilibrium of forces is equivalent to equilibrium of moments.

The approach through a minimal principle rests on entirely different ideas. A potential-energy function is introduced a priori, and the total potential energy of a curve is calculated by integration. By definition, the figure of equilibrium is that which, among all curves of given length connecting two points renders the potential energy a minimum. Exactly one such curve exists, and it is the catenary.

Exercise 29.2. Verify the foregoing outline by carrying out the steps.

The figures obtained by the two methods are the same. If we seek to compare the forces acting, we cannot do so without further assumptions, since in the second method, forces are not used. We may, if we please, define forces from the figure found. If we apply the ideas used in the first method to the figures considered in the second, we find we must say that in the class of figures considered, the principles of statics are not satisfied. The figure of equilibrium, in which, of course, the principles of statics are satisfied, is thus compared with purely hypothetical figures which, according to mechanics, can only be imagined, never produced. Thus often members of a parametric family of comparison curves $y = f(x,\alpha)$ are said to differ from the figure of equilibrium $y = f(x)$ by a "virtual displacement," and sometimes α is thought of as the time in a "virtual process" (of course generally inconsistent with the laws of mechanics), and $\partial f/\partial\alpha$ is called the "virtual velocity."

I mention these old terms because they may have been heard before. While use of them is not wrong, it seems to me to lend itself to the muddy talk that is only too common in "applied" mechanics and in fact to make everything harder rather than easier to understand, so I will avoid them.

We shall see in this lecture and the next how to formulate minimal principles for the equilibrium of simple materials, first without taking effects of different temperatures into account explicitly, and then later including them along with purely mechanical influences.

Stability in Energy.

In a hyperelastic material it is natural to regard the volume integral of the stored-energy function as a total energy in some sense and to explore the consequences of assuming that in a configuration of equilibrium, this energy be a minimum. A little experience suggests that the answer known to be right from direct reasoning will not come out unless the "variational" principle is adjusted somewhat.

Let us suppose given a stored-energy function $\sigma(\tilde{F})$ and a potential-energy function $\upsilon(\tilde{x})$. Let $\mathcal{B}_{\tilde{\kappa}}$ be a region, let $\partial\mathcal{B}_{\tilde{\kappa}}$ be its boundary, let a vector field $t_0(\tilde{X})$ be given upon a part of $\partial\mathcal{B}_{\tilde{\kappa}}$ called S_2, and let the rest of $\partial\mathcal{B}_{\tilde{\kappa}}$ be called S_1. If $\tilde{\chi}$ is any configuration of $\mathcal{B}_{\tilde{\kappa}}$, set

$$S(\tilde{\chi}) \equiv \int_{\mathcal{B}_{\tilde{\kappa}}} [\sigma(\tilde{F}) + \upsilon(\tilde{\chi})]dm - \int_{S_2} (\tilde{\chi} - \tilde{X})\cdot\tilde{t}_o ds \quad . \tag{29.1}$$

A particular configuration $\tilde{\chi}$ is said to be stable in energy with respect to the boundary conditions

$$\underset{\sim}{\chi} = \underset{\sim}{\chi}_o(\underset{\sim}{X}) \quad \text{if} \quad \underset{\sim}{X} \,\varepsilon S_1 \,,$$

$$\rho_{\underset{\sim}{\kappa}} \partial_{\underset{\sim}{F}} \sigma(\underset{\sim}{F}) \underset{\sim}{n}_{\underset{\sim}{\kappa}} = \underset{\sim}{t}_o(\underset{\sim}{X}) \quad \text{if} \quad \underset{\sim}{X} \,\varepsilon S_2 \,, \tag{29.2}$$

if

$$S(\underset{\sim}{\chi}^*) \geqq S(\underset{\sim}{\chi}) \tag{29.3}$$

for every $\underset{\sim}{\chi}^*$ such that

$$\underset{\sim}{\chi}^* = \underset{\sim}{\chi}_o(\underset{\sim}{X}) \quad \text{if} \quad \underset{\sim}{\chi} \varepsilon S_1 \,, \tag{29.4}$$

$$\rho_{\underset{\sim}{\kappa}} \partial_{\underset{\sim}{F}} \sigma(\underset{\sim}{F}^*) \underset{\sim}{n}_{\underset{\sim}{\kappa}} = \underset{\sim}{t}_o(\underset{\sim}{X}) \quad \text{if} \quad \underset{\sim}{X} \,\varepsilon S_2 \,.$$

The concept of stability in energy for the problem of place in finite deformations was first formulated by HADAMARD; the present more general statement is due to NOLL.

$S(\underset{\sim}{\chi})$ is defined by (29.1) is a kind of total enthalpy in the sense of thermostatics. The condition (29.3) asserts that this enthalpy has a weak minimum at a stable configuration. It will turn out, as we shall presently prove, that any minimizing configuration satisfies in $\beta_{\underset{\sim}{k}}$ the same differential equations as does the deformation of a hyperelastic body with stored-energy function σ , subject to body force with potential υ . This being so, $(29.2)_2$ asserts that $\underset{\sim}{t}_0$ is the traction $\underset{\sim}{t}_{\underset{\sim}{\kappa}}$ on S_2, and evidently $(29.2)_1$ asserts that the places into which points on S_1 are mapped by χ be prescribed. Accordingly, a minimizing configuration will satisfy the conditions determining the mixed boundary-value problem of hyperelasticity, and the variational principle (29.3) affords a different statement of it. As is usual with true minimal principles, the configurations $\underset{\sim}{\chi}^*$ compared with the desired one do not satisfy all the same conditions. While it is customary to interpret $(29.4)_2$ as requiring that $\underset{\sim}{\chi}^*$ and $\underset{\sim}{\chi}$ give rise to the same traction $\underset{\sim}{t}_0$ on S_2, we have to stretch a point in order to go along here, since $\underset{\sim}{\chi}^*$ generally will fail to be a possible configuration of the hyperelastic body defined by σ in $B_{\underset{\sim}{\kappa}}$ unless special body forces be brought to bear, certainly different from these corresponding to the potential υ. Indeed, an interpretation in mentioning forces in any way steps outside the realm of the minimal principle and obscures that independence of thought which affords really the only tenable ground in defense of variational principles at all.

We return to the principle (29.3) and work out its consequences.

The Classical Variational Principle.

Consider a parametric family of deformations $\chi^*(X, \alpha)$ such that $\chi^*(X, 0) = \chi(X)$, the desired minimizing configuration. If f is any function of α, and possibly other variables, set

$$\delta f \equiv \partial_\alpha f \Big|_{\alpha=0} . \tag{29.5}$$

Hence

$$\delta F = \nabla \delta \chi . \tag{29.6}$$

where, as usual, $\delta \chi$ is written for $\delta \chi^*$. Also $\delta \rho_\kappa = 0$.

Since S as defined for χ^* by (29.1) is a function of α, a <u>necessary</u> condition for the minimum (29.3) is

$$\delta S = 0 , \tag{29.7}$$

for every family $\chi^*(X,\alpha)$.

By (29.1), (29.7), and (29.6)

$$\begin{aligned}
\delta S &= \int_{B_\kappa} \{[\mathrm{tr}(\partial_F \sigma(F)(\delta F)^T) + (\partial_x \upsilon)\cdot\delta\chi\}dm \\
&\qquad - \int_{S_2} \delta\chi \cdot t_0 ds, \qquad (29.8)\\
&= \int_{B_\kappa} \{\mathrm{Div}[\rho_\kappa(\partial_F\sigma(F))^T\delta\chi] \\
&\qquad - [\mathrm{Div}\,(\rho_\kappa\partial_F\sigma(F)) - \rho_\kappa\partial_x\upsilon]\cdot\delta\chi\}dv - \int_{S_2} t_o\cdot\delta\chi ds, \\
&= -\int_{B_\kappa} [\mathrm{Div}\,(\rho_\kappa\partial_F\sigma(F)) - \rho_\kappa\partial_x\upsilon]\cdot\delta\chi\, dv + \int_{S_1} \rho_\kappa\partial_F\sigma(F)n_\kappa\cdot\delta\chi ds ,
\end{aligned}$$

where $(29.2)_2$ has been used. On S_1, $\delta\chi = 0$ by $(29.2)_1$. In order that $\delta S = 0$ for arbitrary $\delta\chi$ in B_κ, it is therefore necessary that in B_κ

$$\text{Div}\,(\rho_\kappa \partial_F \sigma(F)) - \rho_\kappa \partial_x \upsilon = 0 , \tag{29.9}$$

and conversely, if (29.9) holds, $\delta S = 0$ for all one-parameter families $\chi^*(X,\alpha)$ of fields that satisfy (29.4). The differential equations (29.9) are those for a hyperelastic material with stored-energy function $\sigma(F)$ in equilibrium subject to body force $b = -\partial_x \upsilon$, as follows from (18.20) and (11.14). For the same material, the boundary conditions (29.2), which here we laid down at the outset, correspond to prescribed place on S_1 and prescribed reference traction t_κ on S_2.

Exercise 29.3. Let a deformation field $x = \chi(X)$ and a symmetric tensor field $\tilde{T} = \tilde{T}(X)$ be defined over B_κ ; let $\gamma(\tilde{T})$ and $\upsilon(x)$ be given functions in B_κ; let S_1 and S_2 be the complete boundary ∂B_κ; let t_0 be given on S_1 and χ_0 given on S_2; and set

$$R(\chi, T) \equiv \int_{B_\kappa} \{\frac{1}{2\rho_\kappa}[\text{tr}(\tilde{T}C) - \gamma(\tilde{T}) + \upsilon(\chi)]dm \tag{29.10}$$

$$- \int_{S_1} (\chi - X)\cdot t_0 ds - \int_{S_2} (\chi - \chi_0)\cdot F\tilde{T}n_\kappa ds ,$$

Prove REISSNER's variational theorem: The condition

$$\delta R = 0 \tag{29.11}$$

for fields χ and $\tilde{T}$ such that

$$\begin{aligned} \chi &= \chi_0 \quad \text{if} \quad X \in S_1 , \\ F\tilde{T}n_\kappa &= t_0 \quad \text{if} \quad X \in S_2 \end{aligned} \tag{29.12}$$

is equivalent to the following conditions in B_κ:

$$\mathrm{Div}\,(\underset{\sim}{F}\underset{\sim}{\tilde{T}}) - \rho_{\kappa}\partial_{\underset{\sim}{x}}\upsilon = 0 \ ,$$

$$\underset{\sim}{C} = \partial_{\underset{\sim}{\tilde{T}}}\gamma(\underset{\sim}{\tilde{T}}) \ . \qquad (29.13)$$

Show that (29.12) and (29.13) are equivalent to the boundary-value problem of prescribed place $\underset{\sim}{\chi}_0$ on S_1 and prescribed traction $\underset{\sim}{t}_{\kappa}$ on S_2 for a hyperelastic material in which

$$\underset{\sim}{T}_{\kappa} = \underset{\sim}{F}\underset{\sim}{\tilde{T}} \ , \quad \underset{\sim}{\tilde{T}} = \underset{\sim}{\mathfrak{t}}(\underset{\sim}{C}) = 2\rho_{\kappa}\partial_{\underset{\sim}{C}}\bar{\sigma}(\underset{\sim}{C}) \ , \qquad (29.14)$$

where $\underset{\sim}{\mathfrak{t}}$ is invertible and

$$\gamma(\underset{\sim}{\tilde{T}}) = \mathrm{tr}[\underset{\sim}{\mathfrak{t}}^{-1}(\underset{\sim}{\tilde{T}})\underset{\sim}{\tilde{T}}] - 2\rho_{\kappa}\bar{\sigma}(\underset{\sim}{\mathfrak{t}}^{-1}(\underset{\sim}{\tilde{T}})) \ . \qquad (29.15)$$

Infinitesimal Stability.

Returning to the minimal principle 29.3), we see that it implies the further necessary condition

$$\delta^2 S \geqq 0 \ . \qquad (29.16)$$

We shall consider only "dead loading" in the sense that the body force $\underset{\sim}{b}(\underset{\sim}{x})$ is adjusted so as to remain constant as α varies. That is, $\delta\underset{\sim}{b} = \underset{\sim}{0}$. Then from (29.8) we find that when $\delta S = 0$

$$\begin{aligned} \delta^2 S &= \int_{\mathcal{B}_{\kappa}} \mathrm{tr}\{(\delta\underset{\sim}{F})^T \partial^2_{\underset{\sim}{F}}\sigma(\underset{\sim}{F})[\delta\underset{\sim}{F}]\}dm, \\ &= \int_{\mathcal{B}_{\kappa}} \mathrm{tr}\{A_0[\delta\underset{\sim}{F}](\delta\underset{\sim}{F})^T\}dv \ , \end{aligned} \qquad (29.17)$$

where the elasticity A_0 is that of a hyperelastic material with stored-energy function σ. That is, by (18.20) and (11.20),

$$\mathsf{A}(\underset{\sim}{F}) = \rho_{\underset{\sim}{\kappa}} \partial^2_{\underset{\sim}{F}} \sigma(\underset{\sim}{F})$$

$$A_{k}{}^{\alpha}{}_{m}{}^{\beta} = \rho_{\underset{\sim}{\kappa}} \partial_{F^k{}_\alpha} \partial_{F^m{}_\beta} \sigma(\underset{\sim}{F}) \ , \qquad (29.18)$$

and the subscript 0 indicates that A is to be evaluated at the $\underset{\sim}{F}$ corresponding to the stable $\underset{\sim}{\chi}$. Comparison of (29.17) with (29.16) and (19.9) yields HADAMARD's result,

$$\text{Stability in energy} \Rightarrow \text{Infinitesimal stability.} \qquad (29.19)$$

Thus the condition of infinitesimal stability, introduced and explored in Lecture 19 on a dynamical basis, is found to be a necessary condition for the purely static stability expressed by the minimal principle (29.3). Of course it is not sufficient.

LECTURE 30: THERMOSTATIC EQUILIBRIUM.

The Nature of Thermostatics.

Thermostatics is the science of equilibria at different temperatures of entropies. In the previous lecture we have considered a mechanical theory of equilibrium, in which various putative configurations are compared, and one is selected. The thermostatics of continua is a similar theory but more complicated in that both deformation and entropy or temperature are varied. Again, a purely static theory must be created. The laws of thermodynamics, which we formulated in Lectures 26 and 27, are not to be used, any more than the laws of dynamics are used in theories of pure statics.

This distinction was seen by GIBBS, and he constructed an elegant and complete thermostatics of fluids. His ideas were generalized to simple materials by COLEMAN & NOLL, whose theory we shall now develop.

The Potential for Dead Loading.

In Lecture 21 we have calculated the virtual work done in carrying a body through a homogeneous deformation process $\underset{\sim}{F}(\alpha)$ from $\underset{\sim}{F}(\alpha_1)$ to $\underset{\sim}{F}(\alpha_2)$, subject to the homogeneous Piola-Kirchhoff stress $\underset{\sim}{T}_{\underset{\sim}{\kappa}}(\alpha)$, namely

$$W_{12} = V(\mathcal{B}_{\underset{\sim}{\kappa}}) \int_{\underset{\sim}{F}(\alpha_1)}^{\underset{\sim}{F}(\alpha_2)} \operatorname{tr}(\underset{\sim}{T}_{\underset{\sim}{\kappa}}^T \, d\underset{\sim}{F}) \; . \tag{21.4}$$

If the $\underset{\sim}{T}_{\underset{\sim}{\kappa}}$ is determined from the constitutive equation of a homogeneous elastic material, it will generally vary with α. However, we are free to imagine a process in which $\underset{\sim}{T}_{\underset{\sim}{\kappa}}$ does not obey any particular constitutive equation. If in such a process we hold $\underset{\sim}{T}_{\underset{\sim}{\kappa}}$ constant, we describe that process as occurring at dead loading, and from (21.4) we obtain

$$W_{12} = V(\mathcal{B}_{\underset{\sim}{\kappa}}) \operatorname{tr}(\underset{\sim}{T}_{\underset{\sim}{\kappa}}^T \underset{\sim}{F}) \Big|_{\alpha_1}^{\alpha_2} \tag{30.1}$$

In such a process, then, the virtual work done per unit mass is $\hat{\pi}(\underset{\sim}{F}(\alpha_1)) - \hat{\pi}(\underset{\sim}{F}(\alpha_2))$, where

$$\hat{\pi}(\underset{\sim}{F}) \equiv -\frac{1}{\rho_{\underset{\sim}{\kappa}}} \operatorname{tr}(\underset{\sim}{T}_{\underset{\sim}{\kappa}}^T \underset{\sim}{F}) \ . \tag{30.2}$$

That is, $\hat{\pi}(\underset{\sim}{F})$ is a <u>potential</u> for the surface tractions in this hypothetical process.

More generally, corresponding to any pair $(\underset{\sim}{F}, \underset{\sim}{T}_{\underset{\sim}{\kappa}})$, we may define $\hat{\pi}$ by (30.2). Generally, of course, it will not be a potential for anything.

The following remarks are intended to motivate the definition we shall now give. The result (30.1) will not be used.

<u>Local Thermostatic Equilibrium of Simple Materials</u>.

We shall consider constitutive equations giving the Cauchy stress $\underset{\sim}{T}$, the specific internal energy ε, and the temperature θ as functions of the deformation $\underset{\sim}{F}$ and the specific entropy η:

$$\begin{aligned} \underset{\sim}{T} &= \mathfrak{g}(\underset{\sim}{F},\eta) \ , \\ \varepsilon &= \hat{\varepsilon}(\underset{\sim}{F},\eta) \ , \\ \theta &= \hat{\theta}(\underset{\sim}{F},\eta) \ . \end{aligned} \tag{30.3}$$

These equations are designed to represent a <u>simple material in equilibrium</u> at the configuration $\underset{\sim}{F}$ and the specific entropy η. If η = const. in space, $(30.3)_1$ reduces to (11.2), the purely mechanical constitutive equation defining elasticity, which is the theory of simple materials in equilibrium when no account is taken of the effects of temperature. The <u>thermomechanical</u> potential $\hat{\zeta}(\underset{\sim}{F},\eta)$ is defined as follows for each fixed $\underset{\sim}{T}_{\underset{\sim}{\kappa}}$ and θ:

$$\begin{aligned} \hat{\zeta}(\underset{\sim}{F},\eta) &\equiv \hat{\varepsilon}(\underset{\sim}{F},\eta) + \hat{\pi}(\underset{\sim}{F}) - \theta\eta \\ &= \hat{\varepsilon}(\underset{\sim}{F},\eta) - \frac{1}{\rho_{\underset{\sim}{\kappa}}} \operatorname{tr}(\underset{\sim}{T}_{\underset{\sim}{\kappa}}^T \underset{\sim}{F}) - \theta\eta \ . \end{aligned} \tag{30.4}$$

By its form the thermomechanical potential $\hat{\zeta}$ suggests the free enthalpy ζ of thermodynamics (Lecture 26), but the situation considered here is a far more complex one.

In COLEMAN & NOLL's theory, a class of variations $\underset{\sim}{F}^*$ and η^* of $\underset{\sim}{F}$ and η are considered. This class is defined as follows:

1. η^* is not restricted.
2. $\underset{\sim}{F}^*$ differs from $\underset{\sim}{F}$ by a pure stretch:

$$\underset{\sim}{F}^* = \underset{\sim}{S}\underset{\sim}{F} , \tag{30.5}$$

where $\underset{\sim}{S}$ is positive-definite and symmetric.

3. The local equilibrium of moments is preserved: $\underset{\sim}{T} = \underset{\sim}{T}^T$.

That is, since $\underset{\sim}{T} = J^{-1}\underset{\sim}{T}_{\underset{\sim}{K}}\underset{\sim}{F}^T$, only those $\underset{\sim}{F}^*$ are allowed for which $\underset{\sim}{T}_{\underset{\sim}{K}}\underset{\sim}{F}^{*T}$ is symmetric. Alternatively,

$$\underset{\sim}{T}_{\underset{\sim}{K}}\underset{\sim}{F}^T\underset{\sim}{S} = \underset{\sim}{S}\underset{\sim}{F}\underset{\sim}{T}^T_{\underset{\sim}{K}} , \tag{30.6}$$

where $\underset{\sim}{S}$ is the positive-definite symmetric tensor occurring in (30.5).

COLEMAN & NOLL define $(\underset{\sim}{F},\eta)$ as being a state of local <u>thermostatic equilibrium</u> for assigned $(\underset{\sim}{T}_{\underset{\sim}{K}},\theta)$ if it minimizes $\hat{\zeta}(\underset{\sim}{F},\eta)$ in the above-delineated class of variations, and they <u>postulate</u> that such a minimum exists.

<u>Properties of the Equilibrium Configuration</u>.

According to the postulate just laid down, the constitutive equations (30.3) must be such as to render the function $\hat{\zeta}(\underset{\sim}{S}\underset{\sim}{F}, \eta^*)$ a minimum at $\underset{\sim}{S} = \underset{\sim}{1}$, $\eta^* = \eta$, provided (30.6) holds. Here $\underset{\sim}{T}_{\underset{\sim}{K}}$ and θ are fixed.

Consider the function

$$\hat{\zeta}(\underset{\sim}{S}\underset{\sim}{F}, \eta^*) = \hat{\varepsilon}(\underset{\sim}{S}\underset{\sim}{F}, \eta^*) - \frac{1}{\rho_{\underset{\sim}{K}}} \operatorname{tr}(\underset{\sim}{S}\underset{\sim}{F}\underset{\sim}{T}^T_{\underset{\sim}{K}}) - \eta^*\theta . \tag{30.7}$$

At a minimum, its partial derivatives with respect to $\underset{\sim}{S}$ and η^* vanish.

<u>Exercise 30.1</u>. Show that the vanishing of the partial derivative of (30.7) with respect to $\underset{\sim}{S}$ at $\underset{\sim}{S} = \underset{\sim}{1}$ is equivalent to

$$\mathrm{tr}\{[\underset{\sim}{F}(\partial_{\underset{\sim}{F}}\hat{\varepsilon})^T - \frac{1}{\rho_{\kappa}}\underset{\sim}{F}\underset{\sim}{T}_{\kappa}^T]\underset{\sim}{D}\} = 0 \qquad (30.8)$$

for all symmetric tensors $\underset{\sim}{D}$. Show that because of (30.6) and the frame-indifference of $\hat{\varepsilon}$, (30.8) is equivalent to

$$\underset{\sim}{T}_{\kappa} = \rho_{\kappa}\partial_{\underset{\sim}{F}}\hat{\varepsilon} \quad . \qquad (30.9)$$

Show that the vanishing of the partial derivative of (30.7) with respect to η^* at $\eta^* = \eta$ is equivalent to

$$\theta = \partial_{\eta}\hat{\varepsilon} \qquad (30.10)$$

The results (30.9) and (30.10) show that the functions $\underset{\sim}{\mathfrak{g}}$ and $\hat{\theta}$ in the assumed constitutive equation (30.3) are not arbitrary. If they are to correspond to local thermostatic equilibrium for given $\underset{\sim}{T}_{\kappa}$ and θ, they must be derivable from the function $\hat{\varepsilon}$ as a <u>thermostatic potential</u>.

This is not all. The condition for a minimum asserts that

$$\hat{\zeta}(\underset{\sim}{S}\underset{\sim}{F}, \eta^*) > \hat{\zeta}(\underset{\sim}{F}, \eta) \qquad (30.11)$$

if $\eta^* \neq \eta$ or $\underset{\sim}{S} \neq \underset{\sim}{1}$, $\underset{\sim}{S}$ being positive-definite and symmetric. In view of (30.9) and (30.10), this inequality holds if and only if

$$\hat{\varepsilon}(\underset{\sim}{S}\underset{\sim}{F}, \eta^*) - \varepsilon(\underset{\sim}{F},\eta)$$
$$- \frac{1}{\rho_{\kappa}} \mathrm{tr}[(\underset{\sim}{S} - \underset{\sim}{1})\underset{\sim}{F}(\partial_{\underset{\sim}{F}}\hat{\varepsilon})^T] - (\eta^* - \eta)\partial_{\eta}\hat{\varepsilon} > 0 \ , \qquad (30.12)$$

provided $\eta^* \neq \eta$ or $\underset{\sim}{S} \neq \underset{\sim}{1}$. This is the <u>COLEMAN-NOLL inequality</u>. If $\eta^* = \eta$, it reduces to the C-N condition (20.18) with $\sigma = \hat{\varepsilon}$.

<u>Exercise 30.2</u>. Assume that (30.3) is invertible for η. Write $\eta = \tilde{\eta}(\underset{\sim}{F},\theta)$, and define the <u>free-energy function</u> $\tilde{\psi}$ as follows:

$$\psi = \tilde{\psi}(\underset{\sim}{F},\theta) \equiv \hat{\varepsilon}(\underset{\sim}{F}, \tilde{\eta}(\underset{\sim}{F},\theta)) - \tilde{\eta}(\underset{\sim}{F},\theta)\theta \quad . \tag{30.13}$$

Prove that necessary and sufficient conditions for local thermostatic equilibrium are

$$\begin{aligned} &\underset{\sim}{T}_{\kappa} = \rho_{\kappa}\partial_{\underset{\sim}{F}}\tilde{\psi} \ , \\ &\eta = - \partial_{\theta}\tilde{\psi} \\ &\tilde{\psi}(\underset{\sim}{S}\underset{\sim}{F},\theta^*) - \tilde{\psi}(\underset{\sim}{F},\theta) - \frac{1}{\rho_{\kappa}} \operatorname{tr}[(\underset{\sim}{S} - \underset{\sim}{1})\underset{\sim}{F}(\partial_{\underset{\sim}{F}}\tilde{\psi})^T] \\ &\qquad - (\theta^* - \theta)\partial_{\theta}\tilde{\psi}(\underset{\sim}{S}\underset{\sim}{F},\theta^*) > 0 \end{aligned} \tag{30.14}$$

if $\underset{\sim}{S} \neq \underset{\sim}{1}$ or $\theta^* \neq \theta$, where again $\underset{\sim}{S}$ is positive-definite and symmetric.

<u>Summary</u>.

The results just shown are formally similar in part to those derived from the thermodynamic theory of quasi-elastic response in Lecture 27, in part to those concerning hyperelastic materials in Lecture 20. In part, they may be summarized as stating that a simple material in local thermostatic equilibrium is hyperelastic in two cases:

1. In homentropic states, <u>i.e.</u>, when η = const.,
2. In homothermal states, <u>i.e.</u>, when θ = const.

The stored-energy function σ is $\hat{\varepsilon}$ in the former case, $\tilde{\psi}$ in the latter, and both satisfy the COLEMAN-NOLL condition. Moreover, $\hat{\varepsilon}$ is a convex function of η.

Indeed, the formulae (30.9)-(30.12) and (30.14) show a good deal more, since they are not restricted to any particular class of deformations. They show that $\hat{\varepsilon}$ and $\tilde{\psi}$ may be used as thermostatic potentials in any deformations satisfying the condition of local thermostatic equilibrium. Hence, in particular, $\underset{\sim}{T}_{\kappa}$ is <u>not</u> generally given by (18.20), <u>viz</u>

$$\underset{\sim}{T}_{\underset{\sim}{\kappa}} = \rho_{\underset{\sim}{\kappa}} \partial_{\underset{\sim}{F}} \sigma(\underset{\sim}{F}) \ , \tag{18.20}$$

since in general the potentials $\hat{\varepsilon}(\underset{\sim}{F},\eta)$ and $\tilde{\psi}(\underset{\sim}{F},\theta)$ are not functions of $\underset{\sim}{F}$ alone. That is, according to the COLEMAN-NOLL theory, an elastic material is proved to be hyperelastic in homothermal and homentropic configurations, but not generally.

As we have seen in Lecture 28, similar distinctions hold in the thermodynamic theory of quasi-elastic response.

Closure.

In this course of lectures, like others I have delivered in several parts of the world in the last few years, I have attempted to present important selections from the recent work on rational continuum mechanics, not a full development of it. Some of these selections refer to particular cases in which the phenomena represented and the results obtained seem close to experience and possibly also to experiment. More of them are excerpts from the connected mathematical theory which now unifies a number of older, previously disjoint and more special theories. My emphasis has not been on rigor in the mathematical sense but rather on logical order, frankness, and simplicity. The parts of mathematics used and the vocabulary may not be immediately familiar to persons whose background lies in physics or engineering, but in fact the analysis is very easy, as in any good theory of natural phenomena, and the necessary apparatus can be worked up in at most a month by anyone who remembers his undergraduate mathematics and is willing to try. Indeed, one of the main advantages of the new approach is that it makes everything far easier, far more direct, far closer to the phenomena than ever before. For more detail, the book called The Non-linear Field Theories of Mechanics, published last year as one of the volumes of the Encyclopedia of Physics, may be consulted.

Remarks on the philosophy of science should be avoided by scientists except when they find themselves pushed into a corner. The best philosophical programs are those that can safely be left to be inferred by the student as he progresses. I permit myself in closure only a few remarks on the aim of rational continuum mechanics.

The rational mechanics of mass-points and rigid bodies was formulated at least 150 years ago; now called "analytical dynamics", it is learned as a matter of course by several kinds of scientists. No one has to be told that it is good, clean mathematics, that it is one of the foundation disciplines for physics, and that it is applicable, on occasion, in the design of hardware. No one claims that it is the only kind of valuable mathematics, that it concerns more than a small part of physics, or that it makes horse sense, tests, and handbooks unnecessary when hardware is designed. Some people are more enthusiastic about it than others; some find little or no need for it in their daily work, be it proving theorems in algebra, splitting up sub-atomic particles, or designing parking lots. Those who do not need it are unlikely to be hostile to it. Those who have never learned it are not in a good position to decide whether or not it be applicable to their problems.

Rational continuum mechanics should be seen in the same light. It is a mathematical theory. It does for materials what analytical mechanics does for undeformable masses. It does not claim to solve problems in algebra, to split sub-atomic particles, or to design parking lots. There is no reason for those who do not need it to be hostile to it. As a mathematical theory, it is unlikely to appeal to those who do not like any mathematical theory and who prefer dial-reading, arithmetic, know-how, and guesswork as an approach to nature. My lectures are not for them, nor are they likely to value continuum mechanics more, rather than less, after hearing or reading my lectures. Continuum mechanics will not help them, and it is no threat to them, for they have always existed and surely always will exist in greater number than mathematicians. They were in business long before EUCLID ever proved a theorem; there are more of them in business than ever before; and no one is trying to put them out of business. Continuum mechanics is designed to serve those who feel need for a mathematics of materials, just as analytical dynamics serves those who feel need for a mathematics of fixed inertias. Continuum mechanics makes material response <u>approachable in concept</u>. It has many applications, some known, more to be found; as a mathematical theory, it implies in principle many theorems yet unconjectured and hence unproved; but those who do not learn it are unlikely to find or prove either.

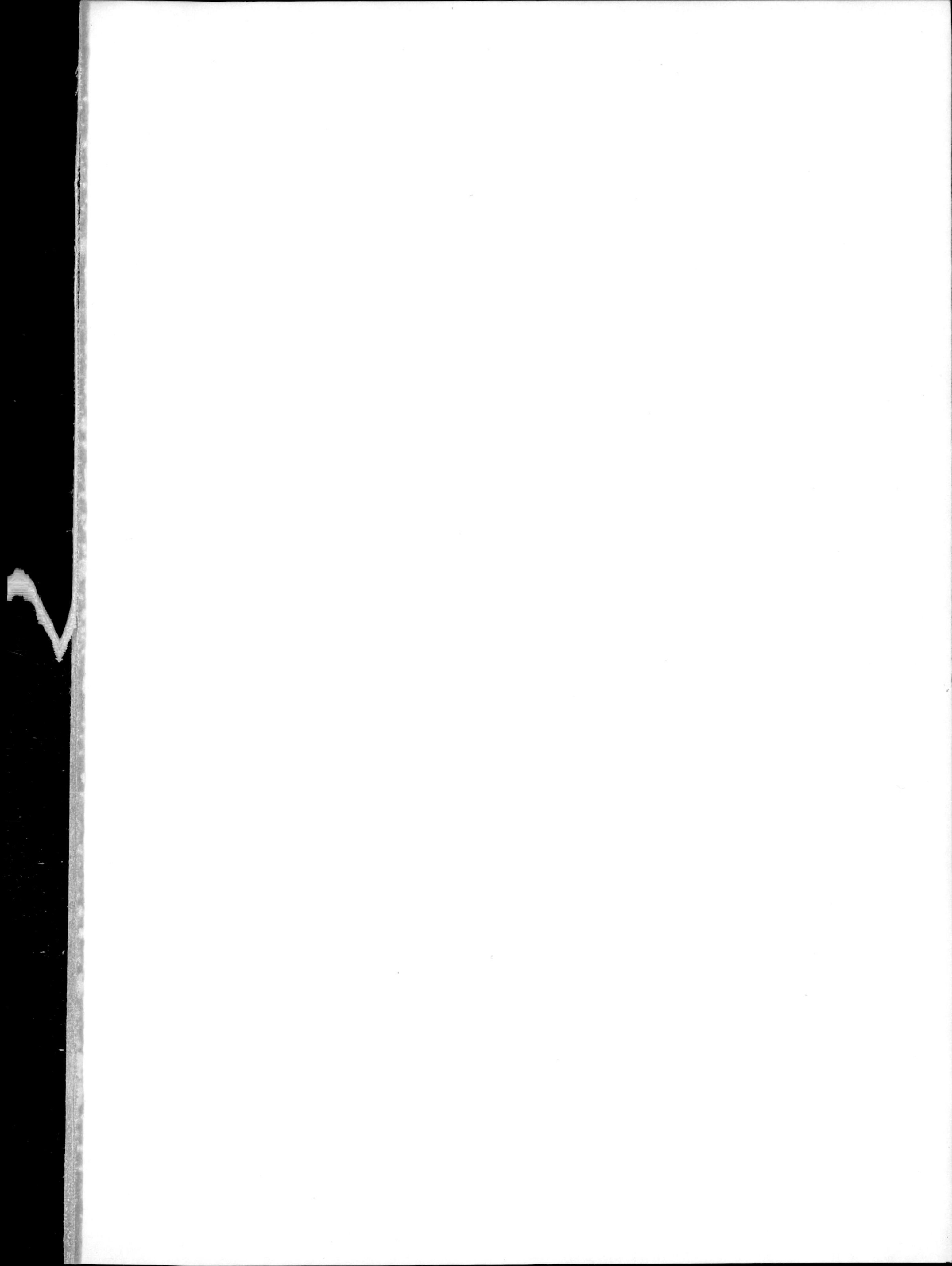